CINE SKILLS AND
ETENCIES

Translational and Applied Precision Medicine
in Clinical Practice

Series Editor: Dhavendra Kumar

GENOMIC MEDICINE SKILLS AND COMPETENCIES

Edited by

Dhavendra Kumar
Honorary Clinical Professor, William Harvey Research Institute,
Bart's and the London School of Medicine & Dentistry,
Queen Mary University of London

ACADEMIC PRESS
An imprint of Elsevier

Academic Press is an imprint of Elsevier
125 London Wall, London EC2Y 5AS, United Kingdom
525 B Street, Suite 1650, San Diego, CA 92101, United States
50 Hampshire Street, 5th Floor, Cambridge, MA 02139, United States
The Boulevard, Langford Lane, Kidlington, Oxford OX5 1GB, United Kingdom

Notices
Knowledge and best practice in this field are constantly changing. As new research and experience broaden
our understanding, changes in research methods, professional practices, or medical treatment may become
necessary.

Practitioners and researchers must always rely on their own experience and knowledge in evaluating and
using any information, methods, compounds, or experiments described herein. In using such information or
methods they should be mindful of their own safety and the safety of others, including parties for whom they
have a professional responsibility.

To the fullest extent of the law, neither the Publisher nor the authors, contributors, or editors, assume any
liability for any injury and/or damage to persons or property as a matter of products liability, negligence or
otherwise, or from any use or operation of any methods, products, instructions, or ideas contained in the
material herein.

ISBN: 978-0-323-98383-9

For information on all Academic Press publications
visit our website at https://www.elsevier.com/books-and-journals

Publisher: Stacy Masucci
Acquisitions Editor: Rafael E Teixeira
Editorial Project Manager: Franchezca A. Cabural
Production Project Manager: Sreejith Viswanathan
Cover designer: Greg Harris

Typeset by STRAIVE, India

" A doctor is a student till his death, when he fails to be a student, he dies"
Sir William Osler
Regius Professor of Medicine, Oxford

To my teachers, trainers, and students

ntents

6. Genomic education based on a shared space for discovery: Lessons from science communication

Jonathan Roberts and Anna Middleton

7. Global perspectives of genomic education and training

Dhavendra Kumar

8. Preparing the workforce for genomic medicine: International challenges and strategies

Desalyn L. Johnson, Bruce R. Korf, Marta Ascurra, Ghada El-Kamah, Karen Fieggen, Beatriz de la Fuente, Saqib Mahmood, Augusto Rojas-Martinez, Ximena Montenegro-Garreaud, Angelica Moresco, Helen Mountain, Nicholas Pachter, Ratna Dua Puri, Victor Raggio, Nilam Thakur, and Rosa Pardo Vargas

9. Digital resources for genomic education and training

Vajira H.W. Dissanayake

Contributors

_. Abadingo Institute of Human _cs, National Institutes of Health, Manila, _lippines

Marta Ascurra National Program for the Prevention of Congenital Defects-Ministry of Public Health and Social Welfare of Paraguay, Asunción, Paraguay

Michelle Bishop Learning and Training Programme, Wellcome Connecting Science, Wellcome Foundation, London, United Kingdom

Kathleen Calzone National Institutes of Health, National Cancer Institute, Center for Cancer Research, Genetics Branch, Bethesda, MD, United States

Eva Maria C. Cutiongco-de la Paz National Institutes of Health, University of the Philippines Manila; National Institutes of Health; Department of Pediatrics, Philippine General Hospital; Health Program of the Philippine Genome Center, Manila, Philippines

Vajira H.W. Dissanayake Department of Anatomy, Genetics, and Biomedical Informatics, Faculty of Medicine, University of Colombo, Sri Lanka

Ghada El-Kamah Human Genetics and Genome Research Institute, National Research Centre, Cairo, Egypt

Karen Fieggen University of Cape Town, Cape Town, South Africa

Clara L. Gaff Melbourne Genomics Health Alliance; Murdoch Children's Research Institute; Department of Paediatrics, The University of Melbourne, Melbourne, VIC, Australia

Desalyn L. Johnson The UAB Heersink School of Medicine, Birmingham, AL, United States

Bruce R. Korf The UAB Heersink School of Medicine, Birmingham, AL, United States

Dhavendra Kumar William Harvey Research Institute, Bart's and The London School of Medicine & Dentistry, Queen Mary University of London, London, United Kingdom

Beatriz de la Fuente Department of Genetics, Faculty of Medicine and Dr. Jose Eleuterio Gonzalez University Hospital, Autonomous University of Nuevo Leon, San Nicolás de los Garza, México

Elly Lynch Melbourne Genomics Health Alliance; Victorian Clinical Genetics Services, Melbourne, VIC, Australia

Ebner Bon G. Maceda Institute of Human Genetics, National Institutes of Health, Manila, Philippines

Saqib Mahmood Department of Human Genetics & Molecular Biology, University of Health Sciences, Lahore, Pakistan

Melissa Martyn Melbourne Genomics Health Alliance; Murdoch Children's Research Institute; Department of Paediatrics, The University of Melbourne, Melbourne, VIC, Australia

Anna Middleton Society and Ethics Research Group, Wellcome Connecting Science; Faculty of Education, University of Cambridge, Cambridge, United Kingdom

Ximena Montenegro-Garreaud Institute of Medical Genetics, Lima, Perú

Angelica Moresco Department of Genetics, J. P. Garrahan National Pediatric Hospital, Buenos Aires, Argentina

Helen Mountain Familial Cancer Program, Genetic Services of Western Australia, Subiaco, WA, Australia

Amy Nisselle Melbourne Genomics Health Alliance; Murdoch Children's Research Institute; Department of Paediatrics, The University of Melbourne, Melbourne, VIC, Australia

Nicholas Pachter Genetic Services of Western Australia, King Edward Memorial Hospital; School of Medicine and Pharmacology, University of Western Australia; School of Medicine, Curtin University, Perth, WA, Australia

Carmencita D. Padilla Institute of Human Genetics, National Institutes of Health, Manila, Philippines

Ratna Dua Puri Institute of Medical Genetics & Genomics, Sir Ganga Ram Hospital, New Delhi, India

Victor Raggio Genetics Department, Faculty of Medicine, University of the Republic, Montevideo, Uruguay

Simon Ramsden Manchester Centre for Genomic Medicine, Manchester, United Kingdom

Jonathan Roberts Society and Ethics Research Group, Wellcome Connecting Science; East of England Genomic Medicine Service, Addenbrookes Hospital, Cambridge, United Kingdom

Augusto Rojas-Martinez School of M and Health Sciences, Tecnologico de M rey, Monterrey, Mexico

Anneke Seller Genomics Education P gramme, Health Education England, Birming ham, United Kingdom

Alison Taylor-Beadling North London Genomic Laboratory Hub, Association of Clinical Genomic Science, Great Ormond Street NHS Foundation Trust, London, United Kingdom

Nilam Thakur National Academy of Medical Sciences, Bir hospital, Kathmandu, Nepal

Emma Tonkin Faculty of Life Sciences and Education, University of South Wales, Pontypridd, Wales, United Kingdom

Rosa Pardo Vargas University of Chile Clinical Hospital, Santiago, RM, Chile

Preface

True education must correspond to the surrounding circumstances or it is not a healthy growth. *Mahatma Gandhi*

Scientific advances, knowledge, and applications are dynamic processes that need to be purposeful and yield benefits to the humankind, environment, and the society at large. Since the completion of the Human Genome Project coupled with rapid advances in OMIC sciences and technologies, practical applications of genomics are now increasingly harnessed. There are several examples around; undoubtedly, the successes of combating and preventing SARS-CoV2 or the COVID-19 pandemic would remain with us for long time. The diagnostic prowess of the reverse transcriptase polymerase chain reaction (RT-PCR) and prevention potential of the ribose nucleic acid (RNA)-derived anti-Corona virus vaccines are now globally appreciated. Every common human, irrespective of skin color, ethnic origin, and geographic location, has become aware of these genomic applications in medicine, health, and socioeconomic welfare.

While we all agree on extraordinary brilliance and aptitude of scientists and technologists leading to new discoveries and innovative applications, we need to remind ourselves that widespread and beneficial applications require special skills and competencies of the frontline workforce. There is no denial on the need for structured and instructive education and training programs to equip the right kind of people trusted for delivering the science and technology. There is an urgent need to create and maintain efficient and effective skilled and competent workforce for delivering the precision genomic medicine and health care.

This book, one of the volumes in the *Genomic and Precision Medicine in Clinical Practice* series, provides theoretical and practical aspects of genomic education and training. The focus is on specific skills and competencies expected from any member of the multidisciplinary team entrusted to deliver the specialized health care. The spectrum is very broad to include clinicians, nurses, counselors, therapists, clinical and laboratory technicians, and community and public health professionals, among others. Each one of these professionals functions within the specific remit and limitations. However, they all must possess core scientific knowledge and practical working skills as expected from a member of the multidisciplinary team. Emphasis is laid on digital medium teaching and virtual simulations for acquiring and enhancing range of skills and competencies. Sections of this book also discuss the ethical and legal aspects of genomic education and training.

In view of the gaps that currently exist between the developed, developing, and underdeveloped nations, sections of the book deal with efforts targeted to fill the gulf between nations. The reader is also reminded about governmental and nongovernmental global programs in genomic education and training, specifically aimed at narrowing the wide gap in skills and competencies among the genomic medicine and healthcare professionals and providers. Needless to say that it would be unfair to assume the finality of information included in the book. The book provides a basic framework and scaffolding on which additional and new information would need to be added. Further, the reader and teachers alike would make the best use of it relevant to specific skills and competencies expected in the genomic healthcare providers.

Dhavendra Kumar, Editor
London, United Kingdom

Acknowledgments

Organizing, compiling, and editing any book require mundane and specialist input from several people. Likewise, this book would not have been possible without the sincere efforts and hard work of the team of world-class experts who used their own time and produced master-class text for this book. It is more commendable that all contributors and their associates worked during the COVID-19/SARS-2 lockdown and recurrent health concerns and welfare issues. I have no words except to salute them for their hard work and dedication exhibited in the final product. Among the world-class contributors, few names deserve special reference to whom I will remain indebted forever, specifically Anna Middleton (Sanger Genome Centre), Angus Clarke (Cardiff University), Vajira Dissanayake (University of Colombo), Simon Ramsden (Manchester Genome Centre), Clara Gaff (Melbourne Genomics), Michelle Bishop (Health Education England), Carmencita Padilla (University of Philippines Manila), and Bruce Korf (University of Alabama).

My close associates within the Global Consortium for Genomic Education (GC4GE) and the Education Committee of the Human Genome Organization International offered invaluable advice and reflections on several occasions. Finally, this book would have been impossible to produce without the professionalism of the Elsevier publishing team. My grateful and humble thanks to the whole team.

I am mindful and fully aware that the modern-day book or any other form of publication is heavily dependent on digital resources. I admit that some of the material appearing in this book might inevitably indicate one or more digital public domains. It should be seen as simply coincidental without any motive to infringe copyrights of any kind. I offer this disclaimer on behalf of all people involved in the production of this educational resource and offer an unconditional apology for any inconvenience caused.

Dhavendra Kumar
March 2022

Foreword

It gives me great pleasure to introduce and commend this new volume, brought together by my most industrious colleague, Professor Dhavendra Kumar. It is the latest in a series of volumes edited by Dhavendra that serve to consolidate the clinical applications of genomics and promote their introduction into new clinical and geographical areas. Extending the scope of genomics into more clinical specialties and more parts of the globe has been Dhavendra's personal mission statement for more than a decade, and this volume contributes very substantially to this task.

Dhavendra and the volume's authors have taken great care to address the educational needs of genetic specialists and nonspecialists so that the needs of clinicians practicing in other specialties are all considered and addressed as well as those of clinical (medically qualified) and laboratory geneticists, genetic counselors, and genetic nurses. The needs of these different groups differ but overlap. Approaching these differences by a focus on competencies is most constructive as it acknowledges that individuals who bring genomics into their clinical work will come from very disparate backgrounds and experiences. For example, a respiratory disease specialist might have conducted molecular genetics research into the population genetics of cystic fibrosis, or they may have worked on the cell biology of the function of different mutant versions of the CFTR protein; either way, they will bring deep knowledge of different aspects of genetics to their work, so their needs for further learning and training will differ from each other and from those of a community midwife or a general practitioner. The focus on competencies allows the recognition of such preexisting knowledge alongside the identification of the individual's needs for further learning given their current clinical role.

The list of contributors to the volume is very impressive. I know or know of them all, personally or professionally or by repute, and can vouch for their ability to cover the topics they have each been set by Dhavendra. While there is a strong representation of the UK among the authors, as is entirely natural, this does not detract from the volume and there are authors from North America, South Asia, Australia, and the Pacific. It is hugely refreshing to see a volume on genomics education that not merely includes a chapter on genomics in developing countries but builds a global perspective into the entire design of the book.

It is also very encouraging to see that so many of the authors are actively involved in the subject matter to be promoted by this book, i.e., in the clinical applications of genomics, so they are well chosen to present a balanced view of the field. It is important for the tensions that arise within genomics to be discussed and for these discussions not to be swept under the metaphorically glossy carpet of "perpetual success" that often resembles corporate gush. There is, first, the tension between what we would all like genomics to deliver and what is scientifically possible, which is often downplayed. And then there is the second tension between what is possible and what is actually being delivered to those afflicted by disease, disability, and deprivation. Both tensions are underrepresented in many of the hyperenthusiastic contributions to public presentations about genomics. In the process of clinical service delivery—as is well known to the clinically active authors—there are many real problems that arise all too often and that straddle the worlds of clinical practice, ethics, public policy, and communication. There is too much of the unwarranted genome hype in much of the literature around specific genome-based projects. I look to this volume to present the problems that arise in clinical practice as well as the promises that may or may not be delivered in reality to the people of our planet.

Angus Clarke
Division of Cancer & Genetics,
The Cardiff University School of Medicine,
All Wales Medical Genomic Service,
Cardiff, United Kingdom

Theories and models for genomics education and training

Melissa Martyn[a,b,c], Amy Nisselle[a,b,c], Elly Lynch[a,d], and Clara L. Gaff[a,b,c]

[a]Melbourne Genomics Health Alliance, Melbourne, VIC, Australia [b]Murdoch Children's Research Institute, Melbourne, VIC, Australia [c]Department of Paediatrics, The University of Melbourne, Melbourne, VIC, Australia [d]Victorian Clinical Genetics Services, Melbourne, VIC, Australia

1 Introduction

The promise of genomic medicine can only be realized if it is incorporated into the practice of clinicians generally, not only the practice of genetic specialists such as clinical geneticists and genetic counselors. To incorporate genomic medicine into practice requires clinicians to change their practice behaviors. The nature of this change will vary, according to the types of genomic applications relevant to that individual's specialty, professional role, and personal level of interest. To support individual change across the workforce, a spectrum of genomic expertise therefore needs to be developed: from increased awareness at a minimum through to the ability to evaluate when genomic applications are ready to be introduced into their specialty and preparedness to lead this change.

It is widely accepted that education plays an important role in developing individual competence in genetics and genomics. Healthcare professionals recognize the importance of genomics education and training to help prepare them to integrate genomics into their practice [1,2]. However, despite intensive genetics education efforts over the past three decades, medical specialists still lack confidence in genomics [3]. This makes sense if we view adoption of genomic medicine by medical specialists as a behavior change challenge, rather than purely an educational challenge. Education needs to be part of a wider strategy to engineer both individual behavioral change and whole of system change, both of which are required to support adoption and incorporation of genomic medicine into clinical practice.

In this chapter, we introduce theories and approaches that can assist genomics educators to design impactful education. We use the term "education" to refer broadly to purposively

Models of education	Setting		Type	
	Educational (external)	Workplace (internal)	Structured	Unstructured

Goals of education	Awareness	Know genomic concepts	Apply to practice	Build networks	Advise others	Lead change

Theory	Learning	Behavior change	Adoption of innovation

FIG. 1   Applying a theoretical approach to educating the workforce in genomics. Theories can inform the design of workforce development approaches that target individual capability and behavior change, as well as the wider system change required for adoption of innovations like genomics. Clarity about the goal of an education activity allows cross-referencing to relevant theories to then identify the educational models most appropriate for achieving particular goals.

designed activities that aim to improve the knowledge, skills, and/or attitudes of the target group. Fig. 1 illustrates concepts that will be introduced in this chapter. Firstly, we outline theories that can underpin design of genomics education. Theories of learning—both cognitive and social—provide insights into how to design an education encounter to achieve optimum learning. Social theories of learning have some similarities with theories of behavior change, both recognizing the importance of the environment in influencing ability to adopt a new practice behavior. Theories of behavior change encourage us to consider how genomics education experiences can be designed to foster development of social networks and practice habits. At the broader system level, theories of adoption of innovation can guide how education can be part of a broader strategy seeking to achieve "whole of system" change. A variety of theoretically informed educational goals can be achieved through different types of education, which may be external to the workplace or internal, and structured or unstructured.

Learning designs are a means to articulate how theory can apply to practice. They can help educators consider the type of change their education program aims to achieve and therefore to select the most appropriate model of learning and specific activities. A program logic approach can also be helpful in planning, developing, delivering, and evaluating genomics education that aligns with theory.

We hope this chapter will assist those in genomics to design and deliver effective education, and those responsible for implementing genomics into clinical practice, to consider how education can be deliberately structured to achieve change.

2 Relevant theories

2.1 Theories of learning

Educational theories describe how learning occurs. There is growing evidence that education interventions that are based on clear theoretical foundations are more effective and have a greater impact on health professional educational outcomes than those without [4,5]. The role of theories of learning in health professional education generally is well established and many excellent summaries of educational theory have been published (see, e.g., [6,7]).

Educational theories may be "grouped" in many ways. We consider first cognitive theories, which concern the internal processes of the learner, and then move to sociocultural theories, which recognize learning as a social process. Key points from these theories of learning are summarized in Table 1, with genetic/genomic examples to illustrate how they can inform approaches to genomics education and training.

TABLE 1   Applying select theories of learning to genomics education and training.

Theory	Key elements	Examples in genomic medicine
Cognitive theories		
Adult learning theory [8,9]	Adult learners • are self-motivated to learn if relevant to professional practice • have varied prior knowledge. Adult learning theory requires that health professionals recognize genomics is relevant to their practice.	A mid-career doctor completes an introductory genomics course as they suspect patients in their practice may benefit from genomic testing.
Situated learning [12]	Use of authentic contexts helps engage learners and encourage reflection and abstraction Situations need to be authentic from the perspective of the learner, e.g., case presentations that focus on the specialty rather than lead with genetics/genomics.	Complex, real-world cases are used for group discussion and reflection in a workshop.
Social theories		
Sociocultural theory [14]	Learning is a social process. Learners "make sense" of new information or experiences by interacting with others to create abstract principles they can apply in other settings. Undesirable learning may be modeled unintentionally, e.g., inappropriate language or attitudes.	A doctor prepares to discuss consent for a genomic test with a patient by • reviewing a "talking points" tool • observing a geneticist undertaking pretest counseling with a patient • practicing with peers. They then take this learning into a supervised session with a patient, eventually conducting pretest counseling unsupervised.
Communities of practice [15]	Learners develop professional expertise and identity through supported but increasingly independent interactions within a community of practice. Learners need to be scaffolded within a community of practice. If language is alienating or tasks too difficult, the learner cannot participate in a legitimate way, leading to reduced self-efficacy and likelihood of incorporating new knowledge and skills into their professional identity.	A nephrologist begins attending a multidisciplinary renal/genetics clinic. Over time, they review more complex genetic cases. Their self-identity expands to include acknowledgment of renal genetics as relevant to their practice.
Coparticipation at work [17]	Informal workplace learning is a combination of workplace opportunities that support learning and individual engagement with those workplace opportunities.	An endocrinologist calls a colleague who is more experienced in genetics for advice about which test to order for a patient with a suspected genetic condition. Over time, the endocrinologist learns which test to order for which type of patient presentation and becomes more independent.

2.1.1 Cognitive theories of learning

Cognitive theories focus on the importance of the learner's internal processes to describe how learning occurs. These theories acknowledge that learners have a range of prior learning experiences and motivation to learn. From a cognitive perspective, learners "make sense" of a learning experience, generating concepts they can then apply across different contexts.

Adult learning theory posits that adult learners are self-motivated due to a recognized need to learn, through interest and/or perceived relevance of educational content to workplace practice. Adult education is recognized as most effective when it is tailored to needs of learners and builds on their experience [8–10]. Adult learners need to apply abstract concepts to their own contexts to consolidate learning (summarized well in [7]). This theory is of course important to models of genomics education, where skilled health professionals require continued education for immediate practical application [6,11]. "Scaffolding" describes the active strategies educators use—such as staging education—to assist learners with progressive levels of complexity as they move through a cognitive learning cycle [6].

Situated learning theory prioritizes authentic learning experiences. Topics are integrated into scenarios that reflect the range of complexity experienced in real-world situations, to apply knowledge and skills to practice, and to motivate and engage learners [12]. Situated learning supports reflection and abstraction, which are critical for enabling learners to make sense of new knowledge and experiences, and transfer learnings to other contexts [10,13]. Situated learning can also increase awareness or interest in a topic (e.g., genomics) if a learner cannot initially see the relevance to their own professional practice [12].

2.1.2 Social theories of learning

Sociocultural theory conceptualizes learning as a social and cultural process, rather than an internal cognitive process [14]. Vygotsky, considered the founder of this theory, argued that for a learner to make meaning from an educational experience, they must interact with others. Learning can therefore be maximized by recognizing the importance of social interactions and incorporating these in the design of educational programs. When considering the apprenticeship models of learning common in health professional training—the "see one, do one" approach—Vygotsky's theory describes the nature of the relationship between the "master" and "apprentice" and how this influences learning, more so than the internal processes of the learner.

Communities of practice theory was influenced by the work of Vygotsky [15]. This theory emphasizes the role of active participation within teams in shaping not only knowledge and skills, but also professional identity. Communities of practice theory stresses the importance of learners participating within a team at a level commensurate with ability, supported to take on more responsibility as skills develop over time. "Scaffolding" in communities of practice focuses less on supporting cognitive processes and more on developing relationships that allow learners to participate at progressively deeper levels within a team over time.

Both theories describe ideal workplace-based learning conditions and why they maximize learning. Over the last three decades, these theories have been applied and extended to improve understanding of structured workplace-based education of health professionals [7]. For instance, studies of medical residency training have identified aspects that enhance and hinder learning (e.g., [16]).

The coparticipation at work theory proposes an interplay between opportunities in the workplace that support learning and individual engagement (the agency of a learner to engage—or not—with those opportunities) [17]. Informal workplace-based learning occurs incidentally through work, without supporting educational structures such as supervision or feedback. Informal workplace learning is less understood than structured workplace learning and theoretical approaches to studying informal workplace-based learning are underdeveloped [18]. For those in practice who have limited capacity to engage in formal education and training, more research is needed to determine how informal workplace learning can be tacitly used to develop genomic competence and confidence within the context of their workplace.

2.2 Behavior change theory

Theories of behavior explain and describe factors underpinning behavior, with models of behavior focusing on explanation in a way that supports development of interventions [19]. For example, the stage of change theory of behavior describes different stages a person passes through on the way to a sustained behavior, with its associated model used to match interventions to the stage of change at which an individual is located [20].

The COM-B model describes behavior (B) as both influencing and being influenced by capability (C), opportunity (O), and motivation (M) [21]. Capability refers to both physical capability—having the knowledge and skills required to perform a certain behavior—and psychological capability, including the belief that one is capable. Opportunity—physical and social—can be understood as the support, or lack of support, offered by a person's environment and network. Motivation is the core of the COM-B model, encompassing both reflective (cognitive) motivation and automatic motivation (habits, instincts, and affect). The COM-B model, widely used to achieve change in health care, was developed specifically to support design and implementation of effective behavior change interventions.

The COM-B model is linked to the Behaviour Change Wheel, a tool to select appropriate behavior change interventions based on the domain being targeted. Fig. 2 illustrates how different types of interventions can be designed to influence capability, opportunity, or motivation, depending on the understanding of what is needed to achieve change for a given individual in relation to genomic medicine. There is good evidence that education is an effective strategy to improve capability [22–24]. However, education can also be used to target other domains in the COM-B model. Table 2 illustrates how educational experiences can be used to create physical and social opportunities and how practical experience through workplace education can impact on reflective and automatic motivation.

2.3 Theories of adoption of innovation in health care

The theories presented earlier focus on individual change. Theories of adoption of innovation address the system-wide changes that enable an innovation like genomics to be adopted and spread within healthcare networks. While education may be only part of a strategy to achieve wider change, these theories can inform the design of education, particularly workplace learning opportunities, to support the desired change.

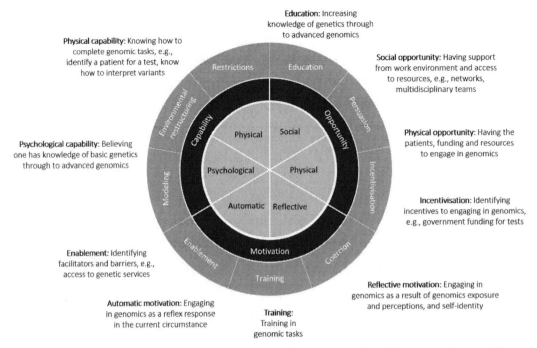

FIG. 2 The Behaviour Change Wheel theoretical framework, adapted from [21,25], encompassing capability, opportunity, and motivation (COM-B), as applied to the practice of genomic medicine. Examples are given of potential behavior change interventions.

TABLE 2 Applying a theory of behavior change to genomics education and training.

Theory	Elements	Examples in genomic medicine
COM-B [21]	*Capability* Both having the knowledge and skills to practice genomics and believing one can *Opportunity* Environmental change alters physical and social opportunities to practice a new behavior *Motivation* Reflective feelings about genomics (positive or negative) and associative habits	A case-based learning workshop on clinical genomics delivered by specialty genomic experts in conjunction with the local genetics service. The workshop directly targets *physical and psychological capability*, by giving specialists knowledge about different types of tests, how to request testing, understand results and apply to patient care. As a result of improved capability (perceived and actual), attendees' *reflective motivation* will also be improved. By introducing the local genetics service and specialty genomic experts, the workshop alters participants' *social opportunity*, as they know who to contact with queries. An immersion experience changes a learner's environment (*physical opportunity*), allowing them to build new social relationships (*social opportunity*). *Automotive motivation* is altered through the practice of genomic medicine, as the learner adopts new habits, imitative practices, and internalizes positive (and negative) experiences and perspectives about the consequences of the behavior. *Physical and psychological capabilities* are improved through workplace practice.

There are a number of theories, models, and frameworks describing adoption, dissemination, and facilitated spread of innovations within healthcare systems (for a summary, see [19]). Different approaches have been proposed to conceptualize and group elements that influence the adoption and sustained use of an innovation. Broadly, these include the innovation itself, the adopter (overlapping with theories of individual behavior change), and the environment in which adoption will be occurring [26].

Considering the innovation itself, complexity is a known barrier to adoption of innovation [27]. Creating an educational environment in which a potential adopter can "trial" the use of an innovation (see, e.g., [28]) reduces complexity and facilitates adoption. Adopters must also be convinced that an innovation is worth adopting, as it offers a "relative advantage" over usual practice [29]. Relative advantage can be conveyed in an educational setting, communicated by peers, or determined through trial of the innovation.

Looking at the environment in which adoption will be occurring, social networks are acknowledged as central in the spread of innovation [29,30]. Education can be designed to foster the development of peer opinion leaders [31] who can champion dissemination of an innovation, such as genomics, and influence development of social networks that support adoption and spread of an innovation. Table 3 illustrates how theories of adoption of innovation in health care can influence design of educational experiences to support use of genomics in practice.

TABLE 3 Applying select theories of change to genomics education and training.

Theory	Key elements	Examples in genomic medicine
Diffusion of innovation [29] Diffusion of innovations in service organizations [30]	Diffusion of innovation may be passive/unintentional (diffusion) or active/planned (dissemination). Social networks are the dominant mechanism for diffusion of innovation <ul><li>Peer opinions influence dissemination</li><li>If key "champions" or "opinion leaders" within an individual's social network support the innovation, they are more likely to adopt it</li><li>Individuals with social ties that span boundaries between the innovation and the organization strengthen dissemination ("boundary spanners").</li></ul> Characteristics of the innovation itself affect its dissemination <ul><li>Complexity and compatibility influence uptake and dissemination</li><li>Trialability improves adoption and assimilation.</li></ul>	Participation in a research project allows a specialist to "trial" genomics, determine relative advantage, reduce complexity, and create new social relationships between the specialty, the genetics department, and the organization. A period of immersion in a genetics service allows development as an "opinion leader." New relationships formed may contribute to the individual acting as a "boundary spanner" who passively diffuses genomics through incidental workplace interactions and actively diffuses genomics by contributing to formal educational efforts. Multidisciplinary meetings may be deliberately constructed to optimize their educational value and to promote relationships between different teams to support active and passive diffusion of genomics in clinical practice.

3 Tools to apply theory to practice

Theories provide a foundation from which educators can develop educational strategies. However, review articles have shown that few genomics education interventions reference theories [23,24]. Learning designs can help *articulate* the specific activities that will be used to achieve an educational goal, and how these activities align with the chosen theory/s. Program logic models can help with the *process* of planning, developing, delivering, and evaluating genomics education, again ensuring that activities and intended outcomes align with the educational goal and theories.

3.1 Learning designs

A learning design is a description of the learning activities used to achieve an educational objective within a theory of learning and is based on what a learner is asked to do [32]. They are useful in mapping how each activity aligns with the goal, theory/s, and model of learning. For example, cognitivist learning designs would typically focus on thinking and mental processes, such as self-directed online modules to provide a foundation in genetics and genomics concepts. In contrast, designs based on social theories of learning would typically include authentic, group case-based activities such as case-review meetings, to support engagement through relevance to practice and collaborative learning with peers.

Learning designs can describe learning at varying levels of granularity, from curricula to subjects or workshops, down to individual activities. For instance, Fig. 3 shows the distinct learning designs of an online self-directed module and a case-based workshop, illustrating the different activities learners complete in each. The learning objectives may be similar, such

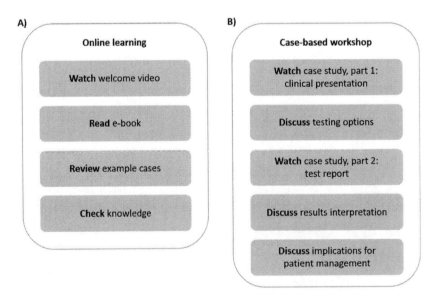

FIG. 3 Learning designs for (A) an online course or (B) a case-based workshop, comprising different types of activities.

as to gain an understanding of the principles and processes of genomic testing as applied to clinical practice, but the two approaches reference different theories. By using verbs to describe the activities, we can see learners are asked to "watch," "read," "review," and "check" in the online learning, versus "watch" and "discuss" in the case-based workshop. The online learning is more informative, aligned with cognitive learning theory, using didactic tasks that involve passive note-taking and individual reflection to provide foundational knowledge. In contrast, the case-based workshop aligns with sociocultural learning theory, using active and peer-based learning activities to build understanding through experience and collaborative reflection.

3.2 A program logic approach

Program logic models are extremely useful when planning, developing, delivering, and evaluating education programs. Logic models highlight the inputs, activities, and intended outcomes of programs [11,33] and can inform evaluation to determine effectiveness and impact [34,35]. Educators can use these models to step through the logic of a program, making explicit the links between chosen theories, overarching goals, and intended outcomes.

As with learning designs, logic models can have different levels of granularity; they can describe education programs or an overarching implementation strategy that may include education as one component. For example, the Genomic Medicine Integrative Research Framework described the implementation of genomic medicine in a "whole of system" logic model spanning context, interventions, processes, and outcomes [35]. Education was just one of the interventions described.

Those providing genomics education do not necessarily have formal education qualifications [36]. A program logic model for genomics education has been developed to support best practice and a theory-informed approach [34]. The model includes prompts to define the overarching goal of an education program and confirm that a need to learn exists [37]. Many have shown the utility of starting with needs assessments when designing genetics/genomics education for health professionals [11,38–41]. Defining the target audience and investigating baseline capability to practice genomic medicine, motivation to learn, and desired level of genomic literacy are essential first steps to inform design of an education program [41]. These insights can help determine which theories and models of learning may be most appropriate to achieve the desired goal and define learning objectives that correspond. The development stage of the logic model also includes prompts for educators to clarify which theory/s will underpin the education program, and how the curriculum and model of learning will align with this through the learning design.

In the context of genomic medicine, the ultimate goal is improved patient outcomes. To achieve this, genomics education should meet immediate learning objectives that align with the chosen goals and theory/s. The immediate outcomes then lead to an intermediate outcome of creating a competent workforce who uses genomic medicine appropriately (long-term outcome). Early in the adoption of genomics within a discipline, the competent workforce includes opinion leaders who actively influence the evidence-based use of genomics by colleagues in their professional networks. The program logic model allows educators to identify suitable measures for each outcome and plan an evaluation program across all stages to assess implementation as well as impact and effectiveness.

4 Models of learning

Genomics education is generally assumed to be structured learning in education settings, such as universities, conferences, or workshops. Structured learning can also occur in the workplace, and incidental, unstructured workplace experiences can also be reframed as learning opportunities.

4.1 Structured learning

Structured learning has a clear goal and aligned learning objectives, with a path (or options) to achieve that goal. Typical examples include university award courses—at undergraduate, graduate, or postgraduate level—and continuing education through workshops or short courses, be they in-person or online. Learning through real-world experience is central to medical education (discussed in [42]); workplace learning may also be structured, if programs have clear goals, with aligned learning objectives and clearly defined structures and resources to attain those goals. For example, medical trainee programs have defined competencies that trainees must prove they have attained by the end of the program. Examples of structured workplace learning to develop genetics/genomics knowledge and skills are emerging in the literature. For example, immersive medical trainee programs in authentic clinical genetic practice settings have been implemented in the United States [43–45] and Australia [46], with participants self-reporting positive impacts on their genetics/genomics knowledge and skills.

To illustrate how different structured models of learning can be used together as part of a strategy to develop the workforce's ability to practice genomic medicine, we provide an example from the Melbourne Genomics Health Alliance program in the Australian state of Victoria (Fig. 4). This multifaceted program used structured models of learning targeted to nongenetic medical practitioners. The program was informed by cognitive and social theories of learning, behavior change, and adoption of innovation in health care. The program offered:

(1) formal continuing education—stand-alone, in-person workshops, or as blended learning whereby online modules were followed by in-person workshops, and
(2) structured workplace learning—a period of "immersion" in a genomics-rich environment, e.g., a genetics service or laboratory [47].

In the Melbourne Genomics program, the goal of the continuing education activities was to provide clinicians with a foundational understanding of genomic technologies, to foster an awareness of the relevance of genomics to their practice, and to create networks centered around peer opinion leaders [31]. The curriculum catered to a range of prior knowledge levels, providing content that was relevant to clinical practice. Case-based learning facilitated by peers with genomics expertise was used to deepen understanding, apply knowledge through small group discussion and reflection, and establish networks that would be useful for clinical practice (see Tables 1 and 2). Learning objectives for the structured continuing education activities were developed that aligned with the chosen theories and aimed to provide nongenetic physicians with basic competence and confidence to implement genomic medicine into clinical practice. The curricula and activities were designed to align with these learning objectives.

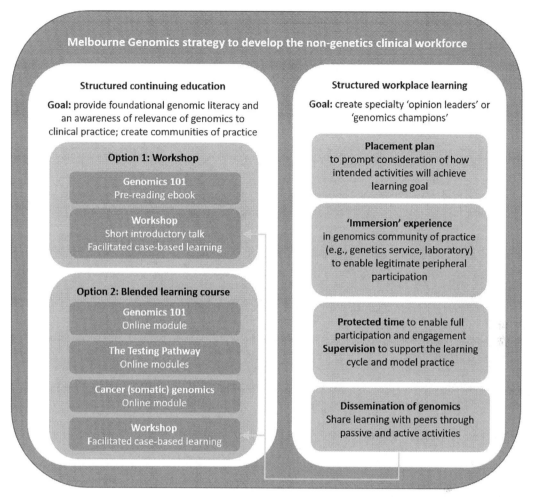

FIG. 4   Example of multifaceted, structured Melbourne Genomics education program, including activities across educational (external) and workplace (internal) settings.

In contrast, the goal of the Melbourne Genomics' structured workplace learning element of the program was to develop a cohort of clinicians who could lead appropriate adoption of genomics within their specialty. To achieve this, the structured workplace learning aimed to improve the capability of participants, support the development of networks and habits that enabled practice change, and promote participants as opinion leaders. Linking the two elements of the program, some immersion participants were asked to develop content for, and facilitate, case-based learning in the continuing education workshops, to provide an opportunity for active dissemination of genomics (see Table 3).

Evaluation of both aspects of the program also aligns with the chosen theories. Longitudinal surveys are assessing changes in knowledge, skills, and behavior as a result of the continuing education, including diffusion activities. Participants who undertook workplace immersion were interviewed and asked to reflect on capability, practice change, recognition by peers, and wider diffusion and dissemination. Interviewees reported that the experience positively impacted both their ability to assess readiness to apply genomics in practice and their actual practice [47]. They reported being recognized as opinion leaders by peers and had under-taken a range of diffusion activities, with some acting as boundary spanners (see Table 3).

4.2 Unstructured workplace learning

Informal workplace-based learning could also be considered continuing education, but lacks structures such as a clear goal, aligned learning objectives, and a deliberate learning de-sign. However, theories of learning can still be used to frame informal workplace experiences as learning opportunities. The coparticipation at work theory [17] draws on both adult learn-ing theory and behavior change theory when describing the dual influences of the existence of workplace learning opportunities, and an individual's engagement with those opportuni-ties (see Tables 1 and 2).

A theoretical lens can help illuminate informal workplace interactions that offer an oppor-tunity for learning [14]. Interviews with Australian physicians from a range of specialties, career stages, and work environments highlighted the value of incidental interactions with colleagues and "genomic experts" for contextualizing knowledge and increasing awareness of the relevance of genomics to their clinical practice [1]. However, the physicians interviewed did not see these interactions as "genomics education" due to their unstructured nature. Examining this experience through the lens of coparticipation at work (see Table 1) suggests these incidental workplace interactions could be used to scaffold learning. Applying a lens of social learning theory to these experiences suggests that providing foundational learning could enable professionals to gain greater benefit from incidental interactions. Conversely, applying a lens of adoption of innovation indicates how structuring education to deliber-ately create specialty opinion leaders could increase the frequency and quality of incidental interactions, thereby generating additional opportunities to learn from expert colleagues and contribute to the diffusion of innovation (see Table 3).

Multidisciplinary teams (MDTs) provide another example of unstructured workplace learning in genomics that can be reframed within sociocultural theory and communities of practice (Table 1). Genomic MDTs involve experts—physicians (geneticists and oth-ers), genetic counselors, and/or laboratory scientists—working collaboratively [48]. The MDT approach provides opportunities for health professionals who are unfamiliar with genomics to become familiar with genetic and genomic concepts in the context of their clinical practice and to facilitate deeper understanding and application of new knowledge and skills into practice. "Scaffolding" may be needed, for example, through participation in foundational education to ensure learning through the workplace experience of MDTs is optimized. MDTs can also foster the development of boundary spanners, that is, peo-ple who bridge areas of practice. Boundary spanners can both informally educate other specialties—bringing their own field of expertise to others at an MDT—and also take new genomics knowledge back to those in their own field and extend the professional networks of their colleagues (see Table 3).

Not all clinicians will be able to access structured workplace learning opportunities in genomics due to practical barriers, such as a lack of infrastructure, opportunity, time, or funding [1]. Also, if the workplace environment is not conducive to applying new learning, then the application of new knowledge and skills to authentic clinical practice may be difficult.

5 Summary

The challenges of learning genomics are not novel. They echo challenges from the genetics era but their relevance is broadening as the clinical applications of genomics expand, impacting more medical specialties [49]. Thinking of the current use of genomics as a continuation of genetics may be reassuring to some and suggests that existing genetics education and training strategies may be transferable to genomics.

Relevant theories of learning, behavior change, and adoption of innovation can be applied to education in a way that specifically seeks to achieve appropriate adoption of genomic medicine. We have described how these theories can underpin models of learning and provided tools to help align theory, learning objectives, and activities through learning designs and program logic.

While structured learning is perhaps most typically considered as education, we also recognize the role of unstructured learning in the workplace. Both structured and unstructured education can be underpinned by theory. In fact, providing both unstructured learning and structured education as part of genomics education and implementation strategies could optimize the likelihood of achieving change. Thinking beyond traditional forms of education allows us to foster additional ways of developing the genomics workforce and assists with implementing genomic medicine into routine health care.

Acknowledgment

Parts of this chapter draw on literature identified in reviews performed by students in our group: Erin Crellin, Emily King, and Alice Kim.

References

[1] B.J. McClaren, et al., Preparing medical specialists for genomic medicine: continuing education should include opportunities for experiential learning, Front. Genet. 11 (2020) 151.
[2] C. Weipert, et al., Physician experiences and understanding of genomic sequencing in oncology, J. Genet. Couns. 27 (2018) 187–196.
[3] E. Crellin, et al., Preparing medical specialists to practice genomic medicine: education an essential part of a broader strategy, Front. Genet. 10 (2019) 789.
[4] R.J. Bernstein, Praxis and Action: Contemporary Philosophies of Human Activity, University of Pennsylvania Press, 2011.
[5] K. Glanz, B.K. Rimer, K. Viswanath, Health Behavior and Health Education: Theory, Research, and Practice, John Wiley & Sons, 2008.
[6] D.C. Taylor, H. Hamdy, Adult learning theories: implications for learning and teaching in medical education: AMEE guide no. 83, Med. Teach. 35 (11) (2013), e1561-72.
[7] S. Yardley, P. Teunissen, T. Dornan, Experiential learning: AMEE guide no. 63, Med. Teach. 34 (2) (2012) e102–e115.
[8] J. Grant, Learning needs assessment: assessing the need, Br. Med. J. 324 (7330) (2002) 156–159.

[9]   K. Knowles, E. Holton, R. Swanson, The Adult Learner: The Definitive Classic in Adult Education and Human Resource Development, eighth ed., Routledge, 2015.

[10]   D.A. Kolb, Experiential Learning: Experience as the Source of Learning and Development, FT Press, 2014.

[11]   C.L. Gaff, et al., A model for the development of genetics education programs for health professionals, Genet. Med. 9 (2007) 451.

[12]   J.S. Brown, A. Collins, P. Duguid, Situated cognition and the culture of learning, Educ. Res. 18 (1) (1989) 32–42.

[13]   D. Schon, The Reflective Practitioner: How Professionals Think in Action, Basic Books, New York, 1983.

[14]   L.S. Vygotsky, The Collected Works of LS Vygotsky: Problems of the Theory and History of Psychology, vol. 3, Springer Science & Business Media, 1997.

[15]   J. Lave, E. Wenger, Situated Learning: Legitimate Peripheral Participation, Cambridge University Press, 1991.

[16]   J.A. White, P. Anderson, Learning by internal medicine residents, J. Gen. Intern. Med. 10 (3) (1995) 126–132.

[17]   S. Billett, Learning through work: workplace affordances and individual engagement, J. Work. Learn. 13 (5) (2001) 209–214.

[18]   L. Evans, Implicit and informal professional development: what it'looks like', how it occurs, and why we need to research it, Prof. Dev. Educ. 45 (1) (2019) 3–16.

[19]   P. Nilsen, Making sense of implementation theories, models and frameworks, Implement. Sci. 10 (2015) 53.

[20]   K. Prochazkova, et al., Teaching a difficult topic using a problem-based concept resembling a computer game: development and evaluation of an e-learning application for medical molecular genetics, BMC Med. Educ. 19 (1) (2019) 390.

[21]   S. Michie, M.M. van Stralen, R. West, The behaviour change wheel: a new method for characterising and designing behaviour change interventions, Implement. Sci. 6 (1) (2011) 42.

[22]   R. Grol, J. Grimshaw, From best evidence to best practice: effective implementation of change in patients' care, Lancet 362 (2003) 1225–1230.

[23]   M. Paneque, et al., A systematic review of interventions to provide genetics education for primary care, BMC Fam. Pract. 17 (2016) 89.

[24]   D. Talwar, et al., Genetics/genomics education for nongenetic health professionals: a systematic literature review, Genet. Med. 19 (7) (2017) 725–732.

[25]   B.J. McClaren, et al., Development of an evidence-based, theory-informed national survey of physician preparedness for genomic medicine and preferences for genomics continuing education, Front. Genet. 11 (2020) 59.

[26]   T. Greenhalgh, et al., Beyond Adoption: a new framework for theorizing and evaluating nonadoption, abandonment, and challenges to the scale-up, spread, and sustainability of health and care technologies, J. Med. Internet Res. 19 (11) (2017) e367.

[27]   H. Bevan, P. Plsek, L. Winstanley, Leading Large Scale Change: A Practical Guide, NHS Institution for Innovation and Improvement, Coventry, 2011.

[28]   C. Gaff, et al., Preparing for genomic medicine: a real world demonstration of health system change, NPJ Genom. Med. 2 (2017) 16.

[29]   E.M. Rogers, Diffusion of Innovations, Free Press, Great Britain, 2003.

[30]   T. Greenhalgh, et al., Diffusion of innovations in service organizations: systematic review and recommendations, Milbank Q. 82 (4) (2004) 581–629.

[31]   L. Locock, et al., Understanding the role of opinion leaders in improving clinical effectiveness, Soc. Sci. Med. 53 (6) (2001) 745–757.

[32]   G. Conole, et al., Mapping pedagogy and tools for effective learning design, Comput. Educ. 43 (2004) 17–33.

[33]   S.C. Funnell, P.J. Rogers, Purposeful program Theory: Effective Use of Theories of Change and Logic Models, vol. 31, John Wiley & Sons, 2011.

[34]   A. Nisselle, et al., Ensuring best practice in genomic education and evaluation: a program logic approach, Front. Genet. 10 (2019) 1057.

[35]   C. Horowitz, et al., The genomic medicine integrative research framework: a conceptual framework for conducting genomic medicine research, Am. J. Hum. Genet. 104 (6) (2019) 1088–1096.

[36]   M. Janinski, et al., Perspectives of Education Providers on Education & Training Needs of Non-Genomic Health Professionals, Australian Genomics, Melbourne, AUS, 2018.

[37]   S.A. Metcalfe, M. Aitken, C.L. Gaff, The importance of program evaluation: how can it be applied to diverse genetics education settings? J. Genet. Couns. 17 (2) (2008) 170–179.

[38]   J.C. Carroll, et al., Informing integration of genomic medicine into primary care: an assessment of current practice, attitudes, and desired resources, Front. Genet. 10 (2019) 1189.

[39] E.J.F. Houwink, et al., Genetic educational needs and the role of genetics in primary care: a focus group study with multiple perspectives, BMC Fam. Pract. 12 (1) (2011) 5.

[40] M. Paneque, et al., Implementing genetic education in primary care: the Gen-Equip programme, J Community Genet. 8 (2) (2017) 147–150.

[41] E.K. Reed, et al., What works in genomics education: outcomes of an evidenced-based instructional model for community-based physicians, Genet. Med. 18 (7) (2016) 737–745.

[42] T. Swanwick, Informal learning in postgraduate medical education: from cognitivism to 'culturism', Med. Educ. 39 (2005) 859–865.

[43] R. Forsyth, et al., A structured genetics rotation for pediatric residents: an important educational opportunity, Genet. Med. 22 (2020) 793–796.

[44] L. Geng, et al., Genomics in medicine: a novel elective rotation for internal medicine residents, Postgrad. Med. J. 95 (1128) (2019) 569–572.

[45] J. Nguyen, et al., Efficacy of a medical genetics rotation during pediatric training, Genet. Med. 18 (2016) 199.

[46] A. Huq, et al., Mainstreaming genomics: training experience of hospital medical officers at the Royal Melbourne Hospital, Intern. Med. J. 51 (2) (2021) 268–271.

[47] M. Martyn, et al., "It's something I've committed to longer term": the impact of an immersion program for physicians on adoption of genomic medicine, Patient Educ. Couns. 104 (3) (2020) 480–488.

[48] S. Bowdin, et al., Recommendations for the integration of genomics into clinical practice, Genet. Med. 18 (2016) 1075–1084.

[49] H. Burton, et al., Genomics in Mainstream Clinical Pathways, PHG Foundation, Cambridge, UK, 2017.

Genetics and genomics education and training in developing countries

Eva Maria C. Cutiongco-de la Paz[a,b,c,d], *Michelle E. Abadingo*[e], *Ebner Bon G. Maceda*[e], *and Carmencita D. Padilla*[e]

[a]National Institutes of Health, University of the Philippines Manila, Manila, Philippines [b]National Institutes of Health, Manila, Philippines [c]Department of Pediatrics, Philippine General Hospital, Manila, Philippines [d]Health Program of the Philippine Genome Center, Manila, Philippines [e]Institute of Human Genetics, National Institutes of Health, Manila, Philippines

1 Introduction

The utility of genetics and genomics in health care has long been recognized. With the increasing availability and affordability of these services, their utilization has also been increasing. The benefit of these services has been realized and has led to its increased demand for clinical geneticists and genetic counselors. The shortage of these professionals has led to an increased need for education of nongenetic healthcare providers, including nongeneticists, nurses, and allied medical professionals [1]. In addition, the rapid advancement and development of genetic and genomic technologies provide an opportunity and also a responsibility of clinical geneticists and genetic counselors to have continuing education in order to provide better and up-to-date information to patients and their families.

Advances in genetics and genomics have stepped up the introduction of genetic services into the healthcare delivery system of the Asia-Pacific region [2]. The Asian continent, which has an area of 43,820,000 km^2, consists of 30% of the world's land area [3]. It is also the most populous continent with an estimated population of 4.6 billion people as of mid-2020 [4]. The region consists mainly of lower middle-income countries (LMICs) and upper middle-income (UMICs) countries [5]. In many LMICs in the region, the emergence of the middle-class population, the lack of regulatory oversight, and weak capacity building in medical genetics expertise and genetic counseling services lead to a range of genetic services of variable quality with minimal ethical oversight [2].

2 Burden of genetic disease

The prevalence of genetic disorders is approximately 5.3%, which translates to about 416 million individuals affected with a genetic condition worldwide. This estimate increases to 7.9% or 621 million individuals worldwide if all congenital anomalies are included. This number further rises if we include other diseases with a known genetic basis or conditions suspected to have a strong genetic basis [6–8]. In the latter, an estimated 369 million in the Asian continent have a condition with a genetic basis with thalassemia being one of the most common single-gene disorders in many Asian countries [9].

Medical genetics services aim to assist patients and families with genetic conditions including metabolic disorders, hereditary cancers, structural or functional birth defects, chromosomal disorders, and complex multifactorial diseases, or those with risks of having these [10,11]. Genetic risk assessments and counseling are rendered.

In developing countries, challenges to increasing access to genetics services have been fundamentally the shortage of trained specialists, the lack of appropriate technology, the scarcity of funding, and the lack of links with the primary healthcare level [12]. Adequacy and acceptance of genetics and genomics services in the developing world are faced with a number of barriers including limitation of resources, genetic conditions not being prioritized, misconceptions that most of these conditions are linked to sophisticated technology and expensive management, and low genetics literacy. In addition, an insufficient number of trained healthcare professionals, inadequate data on the exact magnitude, and the economic burden of genetic and congenital disorders are some factors to be considered in the developing countries [13].

3 Academic programs in genetics and genomics

3.1 Subspeciality training in clinical genetics and genomics

Medical or clinical geneticists are physicians who specialize in the branch of medicine associated with the interplay between genes and health. They are trained to examine, diagnose, manage, and counsel individuals with hereditary disorders as well as their families. The principles and practice of genetics and genomics essential to the education of trainees in this field include human genetics and genomics, cytogenetics, normal morphology and dysmorphology, embryology, fetal pathology concerning birth defects, normal human growth and development, teratology, biochemical, and molecular genetics. It also includes learning the approach to the diagnostic recognition of patterns of malformation, the method in looking for morphologic clues to assist in the timing and designation of pathogenesis to structural defects, the logical way to handle a case of inborn error of metabolism, the knowledge of the natural history of common malformation syndromes and inborn errors of metabolism, the management of prenatal genetic conditions and adult-onset hereditary diseases, and the application of the principles of genetic and genomic epidemiology [8].

In the study by Sirisena and Dissanayake [14], strategies for establishing training in core competencies of genetics and genomics at the undergraduate, postgraduate, and continuing professional levels need to be in place in order to implement genomic medicine at different levels of healthcare delivery. A gap in knowledge has been identified across the board among healthcare professionals, including medical graduates, general practitioners, nurses, and nongenetic specialists [15,16]. Assessment of the currently available genomics professional workforce and estimates of workforce needs will help to prioritize educational programs that can be tailored to the settings in which genomics-based care will be delivered [16].

In many developing countries in Asia, education and training in clinical genetics and genomics are in different stages and levels of development and maturity. Table 1 shows a general overview of the training programs in some developing countries in Asia.

3.1.1 India

As of 2020, there are 83 trained medical and clinical geneticists practicing all over the country. Many of the pioneering geneticists have received training abroad. However, the establishment of the medical genetics residency program at the Sanjay Gandhi Post Graduate Institute of Medical Sciences (SGPGIMS), Lucknow, more than 30 years now has been the primary training source for the newer generation of geneticists. At present, the Medical Council of India and National Board of Education have 3-year "superspeciality" training programs in medical genetics both in DM (Doctor of Medicine) and DNB (Diplomate in National Board). In addition, many trained geneticists practice clinical genetics as part of their pediatric practice [17]. A research thesis is mandatory for completion of training.

3.1.2 Indonesia

Most of the physicians working as clinical geneticists in Indonesia have trained overseas for their PhD degrees in medical genetics. Training on clinical genetics started in 1994 primarily focused on specific areas of expertise for many medical genetic centers in Java. This includes metabolic genetic diseases, cytogenetics, intellectual disability, disorder of sexual development, hemoglobinopathies, hereditary cancer, genome diversity, and mitochondrial diseases [8].

The Eijkman Institute for Molecular Biology, one of the most prominent research institutions in Indonesia for medical molecular biology and biotechnology, provides genetics training mainly in advanced laboratory work for the diagnosis of genetic diseases through a genetics clinic manned by genetic counselors. Since 2001, the Center for Biomedical Research, Faculty of Medicine Diponegoro University, is in collaboration with the Radboud University Nijmegen Medical Center in the Netherlands in organizing the Annual Medical Genetics Course, a short course in genetics from basic to the clinics [8].

3.1.3 Malaysia

Clinical geneticists in Malaysia were initially trained in Australia, the United States of America, and the United Kingdom. All these pioneering medical geneticists in the country obtained their speciality qualifications in pediatrics before proceeding to subspeciality training in clinical genetics [18].

The subspeciality of clinical genetics was recognized in 2005 in Malaysia. All specialists practicing clinical genetics and metabolic medicine who fulfilled the "grandfather clause" were accepted as clinical geneticists under this subspeciality in pediatrics. The duration of training is 3 years, with at least 1 year of training abroad in a recognized center for clinical genetics or metabolic medicine. The training program consists of five core competencies, namely clinical skills competencies, specialized knowledge, technological skills, management skills, and academic skills. The training is done in the form of clinic and on-call duties, continuing medical education activities, genetic counseling sessions, ward rounds, case discussion, case conferences, seminars, journal sessions and teaching and supervision of junior doctors, nurses and other allied staff, and involvement in research over the duration of 3 years [8].

TABLE 1 General overview of the clinical genetics and genomics education, training, and service in developing countries.

	India	Indonesia	Malaysia	Philippines	Sri Lanka	Taiwan	Thailand	Vietnam
Estimated total population	1.39 Billion	275.7 Million	32.6 Million	110.7 Million	21.4 Million	23.8 Million	69.9 Million	98 Million
Clinical geneticist to population ratio	1:16.7 Million	1:39.4 Million	1:2.3 Million	1:6.5 Million	1:1.3 Million	1:391,000	1: 3 Million	1:6.8 Million
Number of clinical geneticists	83	7	14	17	>16	66	25	>13
With local training program	YES	YES	YES	YES	YES	YES	YES	YES
Year that clinical genetics training program started	1990—DM Medical Genetics 2015—DNB Medical Genetics	1994	2000	2000	2010	1990	2013	2017
Existence of a national regulatory body for clinical geneticists	NO	NO	YES	NO	NO	YES	YES	YES

Also part of the training is a community immersion where trainees work with nongovernmental organizations and support groups in the community.

During the training, the trainee will keep a record of cases managed under supervision in a logbook. Majority of the cases will be seen in the pediatric genetic clinics. Trainees are also recommended to attend a 3-month rotation at the familial cancer genetics clinics; feto-maternal medicine clinics; and other specialized genetics clinics that may be available such as skeletal dysplasia clinics and cranio-maxillofacial clinics [8,19]. In addition, the trainees are also required to spend time in genetic laboratories with a record of involvement as either observation or analysis of cases during their posting.

3.1.4 Philippines

In 2000, a Philippine Pediatric Society-accredited Clinical Genetics Fellowship Program was established. The 2-year program is designed to provide the knowledge, understanding, and skills required for the competent diagnostic evaluation, management, and genetic counseling of patients with genetic disorders and their families [8,20].

The training includes performance of a careful, detailed physical examination to ascertain normal and abnormal morphogenesis, detection and distinction of major anomalies, minor anomalies and normal variants, collection of a carefully detailed history with respect to both the prenatal and postnatal periods including pedigree construction and interpretation, come up a differential diagnosis of single primary defects or multiple malformation syndromes, differential diagnosis of possible inborn errors of metabolism, critical analysis of the medical literature and computerized databases, diagnostic synthesis, longitudinal follow-up of patients, genetic counseling of patients and families, and organizing multidisciplinary clinics. The following regular activities are part of the fellowship training program: perform daily rounds on in-patient referrals/admissions, clinical, and metabolic genetics clinic twice weekly; participate in the fellows learning encounters which include case reviews, lectures by consultants, and the metabolic hour; administer enzyme replacement therapy; assist the newborn screening program in the diagnosis and management of conditions in the expanded newborn screening panel; and rotate in the different laboratories such as cytogenetics, biochemical genetics, molecular genetics, and newborn screening [8].

With the increasing awareness of the significant contribution of genetics to various diseases, the Clinical Genetics Fellowship Program has expanded to providing prenatal counseling services for pregnancies at high risk for birth defects as well as risk assessment and genetic counseling for patients with cancer and their families. To increase accessibility to genetics clinical services to geographically isolated and disadvantaged areas of the country, the fellows are also trained to participate in providing telegenetics services [8].

At the end of the training period, the trainee should have a solid knowledge of the clinical approach to patients with suspected syndrome diagnosis and suspected metabolic disorders. The trainee must be competent in the interpretation of genetic diagnostic examinations and provide management as needed. The trainee also must accomplish a completed research and is encouraged to write case reports/case series of interesting cases [8].

3.1.5 Sri Lanka

The Human Genetics Unit (HGU) in the Faculty of Medicine, University of Colombo, was established in 1983. In 2010, Masters Courses in Clinical Genetics and Genetic Diagnostics was started in collaboration with the University of Oslo, Norway. A master's course in stem cell biology and regenerative medicine was established in 2012 in collaboration with the Manipal University, India. As of 2019, these master's courses have produced 16 clinical geneticists and 21 scientists [21]. In 2020, Genetics became a subspeciality in Medicine under the specialities of pediatrics, medicine, and obstetrics and gynecology. Interested doctors under these specialities may have further training in genetics and genomics after their senior registrar period. The training program is for two years, one year in Sri Lanka and one year abroad.

3.1.6 Taiwan

The clinical genetics training program in Taiwan includes clinical services, seminars, journal reading, case conferences, genetic counseling, and continuing medical education activities. Twenty-five genetic cases as well as twenty-five metabolic cases must be reported by the trainee with supervision by a qualified training doctor, to apply for the genetics board examination. Additionally, the trainees are required to spend 3 months each training course in biochemical genetics, molecular genetics, and cytogenetics laboratories. The total training duration is a 2-year full-time training. Research engagement is required and is included in the 2-year fellow training course. It is required for the trainee to publish at least one original article before getting the certification of genetics and metabolism subspeciality.

Qualified training centers are required to have both clinical services and laboratory facilities in genetics and metabolism. The ratio of the number of trainees per trainer is suggested not to be over two. By 2020, there are already 11 qualified training centers for genetics and metabolism subspeciality in Taiwan, including four in northern Taiwan, two in middle Taiwan, and five in southern Taiwan. Fetal-maternal medicine specialists provide prenatal diagnosis services, with the collaboration with the medical geneticists and genetic counselors. The genetic counseling centers are certified by the government. The laboratories providing testing for a certain genetic diagnosis need to be certificated by the government.

3.1.7 Thailand

In Thailand, the formal fellowship training program in clinical genetics was established in 2013 by the Royal College of Pediatrics under The Medical Council of Thailand. Internists and pediatricians with a special interest in this field may subspecialize in clinical genetics. The training program follows a competency-based curriculum. Patient care training is accomplished by seeing patients in the clinic and in the hospital. Knowledge and skills are enhanced through activities such as genetic counseling sessions, case conferences, seminars, and journal sessions. Practice-based learning and improvement in care of patients are developed through patient encounters in both the outpatient and in-patient departments of the hospital. Interpersonal relationships, communication skills, and professionalism are also enhanced during these encounters. Genetic counseling sessions and teaching and supervision of junior doctors, nurses, and other allied staff are also part of the training activities [8]. Presently, telemedicine including telegenetics has been introduced to the trainees.

In addition to the experiences gained at the Department of Pediatrics and Internal Medicine, the trainee is expected to do a 3-month posting at the cancer genetics clinics, fetomaternal medicine clinics, and other specialized genetics clinics such as orthopedics and surgical clinics. The trainees are also required to spend 4-month rotations in cytogenetics, molecular genetics, and biochemical genetics laboratories. Trainees are also allowed to have a 2-month period for an elective area of their interest such as hematology or neurology [8].

Research in medical genetics is one of the main thrusts of the training program. The trainee has to complete one research within the 2 years of training and is expected to publish it in a peer-reviewed international journal. After their second year, trainees apply for the medical genetics subspeciality board examination both in written and practice. Once they pass the examination and all the requirements stated in the curriculum are completed, they are registered as a medical geneticist with the Thailand Medical Council [8].

3.1.8 Vietnam

In 2018, there are about 13 trained medical geneticists in Vietnam [22]. It has been increasing since then, along with the expansion of public and private genetic testing laboratories and services. Most of the geneticists in Vietnam are trained abroad with master's and PhD degrees, and fellowship training. For 4 years now, postgraduate programs in clinical genetics have been offered to residents and master's course students.

3.2 Genetics and genomics in the medical school curriculum

Genetics and genomics are becoming an important part of medicine. The physician must be able to understand and communicate complex information to patients in a simple and accurate manner [23]. Integration of genetics and genomics into the medical school curriculum is necessary. Ideally, the inclusion of genetics and genomics in the medical curriculum can bridge the gap between the basic sciences and clinical training [14,16].

In Taiwan, medical genetics content is included in the medical ethics and other specialist classes of medical students. In addition, medical students also have opportunities to have their rotation/clerkship/internship in medical genetics in many universities overseas. In Indonesia, clinical genetics is taught as part of the human genetics course in the undergraduate medical schools of state universities. Medical schools in Malaysia have a genetics syllabus in the preclinical undergraduate medical program using problem-based learning and lectures. This is further supported by case-based teaching of clinical genetics and metabolism in the ward rounds, tutorials, and seminars. In Thailand, genetics and genomic medicine are included in the medical school curriculum, as well as in pharmacy, dental medicine, medical technology, and nursing schools. In the Philippines, clinical genetics modules are taught as part of the syllabus for undergraduates in medical school [8]. During their clinical clerkship and internship, medical students also have the opportunity to attend the clinical genetics outpatient clinics as part of their rotation in pediatrics at the Philippine General Hospital. The Medical Council of India has been working on upgrading the genetics curricula in the undergraduate and postgraduate courses [17]. Currently, basic genetics is integrated into the anatomy module and into the pediatrics and general medicine rotations. In Sri Lanka, medical genetics is part of the curriculum in surgery, medicine, obstetrics and gynecology, pediatrics, family medicine, transfusion medicine, hematology, oncology and radiotherapy, molecular medicine, community dentistry, legal medicine and forensic science, medical administration, and sports medicine [21]. In Vietnam, there are ten medical schools that offer genetics lectures. This includes 45 h of medical genetics lectures for 4th-year medical students like genetic counseling and 40 h of medical genetics for postgraduate courses.

3.3 Training programs for nongenetic healthcare workers

The specific barriers to increasing the equitable provision of genetic services in developing countries vary according to the local community, region, and country. The lack of trained healthcare personnel is one of the identified impediments in the delivery of genetics and genomics services. This situation is common in developing countries [12]. There is also a high clinical geneticist to country population ratio. This is addressed by training programs for nongenetic healthcare workers.

In Indonesia, clinicians from any department are invited to an annual medical genetics course. In the postgraduate pediatric program in Malaysia, clinical genetics is a recognized subspeciality and accorded dedicated slots for teaching, intensive courses, and examinations. In India, the Department of Medical Genetics at the SGPGIMS holds a two-week training for clinicians, medical college teachers, and private practitioners. The training gives an overview of basic genetics and approach to common genetics problems using diagnostics and genetic counseling [17]. In Sri Lanka, continuing medical education courses in genetics and genomics for medical and allied healthcare professionals are regularly held. A 3-month-long certificate course in genetics is offered to medical doctors. Raising genetic awareness among the public is also being conducted in collaboration with medical colleges and associations in the country [14]. In the Philippines, the Institute of Human Genetics, National Institutes of Health, conducts various training activities and webinars for nurses, dietitians, physicians of other subspecialities, and other healthcare workers. These include topics on prenatal genetics, cancer genetics, and topics on disorders in the newborn screening. In Taiwan, societies, especially for pathologists and for oncologists, also provide training programs (usually 2–3 days) discussing cancer genetics. In Vietnam, clinical genetics is currently included in master's programs and residential courses in 14 medical schools.

3.4 Genetic counseling

Genetic counselors (GCs) are healthcare providers who have specialized training in medical genetics and psychosocial counseling of patients and family members with risks for genetic disorders. As of 2018, there are an estimated 350 genetic counselors in the Asian continent. The development of genetic counseling training has been diverse as the profession becomes established [24].

Table 2 shows a general overview of genetic counseling education and training in some developing countries.

3.4.1 India

The first program in genetic counseling offered locally was established in 2007 by Kamineni Hospital, Hyderabad. It is a 1-year Postgraduate Certificate Course in Medical and Genetic Counseling. The second program is a 1-year Certificate Course in Genetic Counseling established in 2012 by Manipal Hospital, Bangalore. The training provided includes a didactic program with clinical placements in various laboratories and departments such as neurology, pediatrics, fetal medicine, and cancer genetics. Students would have ideally completed a bachelor's or master's degree in genetics, biological sciences, nursing, psychology, or medicine. The latest program offered is a 2-year master's program at Kasturba Medical College,

TABLE 2 General overview of the genetic counseling training and service in developing countries.

	India	Indonesia	Malaysia	Philippines	Sri Lanka	Taiwan	Thailand	Vietnam
Number of genetic counselors (GCs)	180	90	9	14	None (clinical geneticists do the genetic counseling)	~ 100	~ 100 healthcare professionals with genetic counseling training for thalassemia only	Not available but there are GCs trained abroad
With training program in genetic counseling (Year started)	YES (2007)	YES (2006)	YES (2015)	YES (2011)	NO	YES (2002)	YES For thalassemia (2005); 6-month course (2021)	NO
Existence of a national regulatory body for genetic counselors	YES	NO	NO	NO	NO	YES	NO	NO
Estimated genetic counselor-to-population ratio	1:7.7 Million	1:3.1 Million	1:3.6 Million	1:7.9 Million	N/A	1: 238,000	N/A	N/A

Manipal [25,26]. Minor and major research projects are required in some courses of the program. A community immersion is also included in the program where students are required to visit special schools. They also participate in the activities of the thalassemia and sickle cell society, advocacy groups, and healthcare camps.

Currently, the Board of Genetic Counseling is the regulatory body that looks after the education, training, and practice of genetic counseling in the country.

3.4.2 Indonesia

The Master of Science program in Genetic Counseling has produced 90 genetic counselors as of 2015. This master's degree program in Indonesia is the very first in Asia. It was established in 2006 at the University of Diponegoro. There have been collaborations with more advanced centers overseas to maintain the quality of this degree program in genetic counseling. Enrollees have mostly been medical doctors, who learned about clinical genetics and genetic counseling through this course and have been very helpful in running the genetics clinics [8]. Students in the program must do genetic counseling practice at special schools to see teachers, parents, and families. Presently, telegenetics has been utilized a lot. It is also a requirement for students to do research and publish before graduation. Most of them have done genetics laboratory work.

3.4.3 Malaysia

Genetic counseling services have been available in Malaysia since 1994 [18]. In September 2015, a postgraduate program for a Master of Science (Genetic Counseling) was established at the National University of Malaysia, with the 1st and 2nd student cohort having two and eleven trainees, respectively. Currently practicing senior clinical geneticists and GCs from various public and private academic institutions have contributed as visiting lecturers.

3.4.4 Philippines

The Master of Science in Genetic Counseling under the Department of Pediatrics, University of the Philippines, Manila, was first offered in 2011. The curriculum was designed with inputs from various disciplines like bioethics, biostatistics, epidemiology, medical anthropology, and public health [27]. It includes clinical and research areas with a significant focus on public health, including rare disorders and newborn screening. In addition to didactics and clinical exposures, student-preceptorship collaborations are arranged with certified genetic counselors in the United States and Canada to further augment the learning of genetic counseling students. It entails regular video case conferences for students and faculty members to share challenging genetic counseling cases ensuring protection of patient confidentiality and privacy [26,27].

Most of the students have prior background in the sciences or various healthcare professions. Education is provided in English and Filipino, taking into account the many dialects across the Philippines. Funding is available for individuals from diverse regions to attend the program and return to their province in order to provide services countrywide [26].

3.4.5 Sri Lanka

Genetic counseling services are also offered in Sri Lanka, which is primarily given by clinical geneticists. Various subspeciality clinics such as dysmorphology, reproductive genetics, cancer genetics, hematology and hemato-oncology, neurogenetics, and ophthalmic genetics provide clinical genetics and genetic counseling services [21]. There are no professional GCs in Sri Lanka.

3.4.6 Taiwan

The Institute of Genome Sciences, National Yang-Ming University, offered training courses in 2002–2004, while the National Taiwan University (NTU) started a master's program on genetic counseling for in-service students in 2003. As of 2021, this program continues to provide didactic education and clinical training for individuals interested in this field. The graduate programs include courses in medical genetics, statistics, counseling theory and skills, cell and molecular biology, prenatal genetics, cytogenetics, embryology, and human psychology. Clinical rotations, a 12-month internship, cover prenatal, postnatal, and general genetics clinics. A rotation in a diagnostic laboratory is required as well. A thesis is required for graduation.

Certification of genetic counselor is currently offered separately by Taiwan Human Genetics Society and the Taiwan Association of Genetic Counseling. Certified GCs are required for a Ministry of Health-accredited genetic counseling center. Genetic Counselor Licensure is not yet available in Taiwan but is currently in the planning process.

3.4.7 Thailand

Genetic counseling education in Thailand is primarily fueled by the needs of the country. The Ministry of Health recognized the need for genetic counselors for the prevention and control of thalassemia, the most common genetic disease in Thailand. Short-course training programs (usually 3 to 5 days long) on genetic counseling for this special purpose have been started in 2005, which has produced almost 100 healthcare personnel equipped to provide the much needed genetic counseling services [8].

The genetic counseling program was recently launched in 2021 in cooperation with councils (medicine, pharmacy, nurse, dentist, and medical technology). The short six-month program is adapted for those who have been working in clinical genetic clinics such as genetic nurses, clinical pharmacists, or bioinformatics. The program is a pilot short course for the certificate of genetic counseling.

3.4.8 Vietnam

A master's degree program in genetic counseling is not yet available in Vietnam. However, there are collaborations with Mary-Claire King's laboratory supporting the development of a genetic counseling program and clinic time in Hanoi, Hue, and other potential genetics clinics in Vietnam [22].

3.5 Genetic and genomic nursing

Genetic and genomic nursing has been a new area of study in nursing science. It is an ethical and philosophical product of nursing science development as an adaptation to global health issues. This need has been recognized in developing countries [28]. Nurses, in general, provide patient education, care coordination, and psychosocial support to patients and their families. They also administer and monitor medications and therapies for patients [1]. Specifically, genetics nurses perform risk assessments, analyze the genetic contribution to disease risk, and discuss the impacts of risks on healthcare management for individuals and families. They also provide genetics education, provide nursing care to patients and families, and conduct research in genetics.

4 Professional societies and patient support groups in genetics and genomics

Professional societies in genetics and genomics play an important role in the education of healthcare providers in genetics and nongenetic healthcare professionals. These specialized groups are organized to facilitate continuing education and update the knowledge of their members and guests. Activities prepared by these groups may also provide best practices and experiences of genetics healthcare professionals. This is very much important, especially in the developing world.

Prior to the inaugural meeting of the Asia Pacific Society of Human Genetics (APSHG) in 2005, an informal group of medical and human geneticists in the region already met regularly. In 2006, it was registered in Singapore and later was accepted as a full member of the International Federation of Human Genetics Societies in Brisbane, Australia, during the 11th International Congress of Human Genetics. The members of the organization include scientists, clinical geneticists, genetic counselors, and students in the Asia-Pacific region. The APSHG aims (a) to promote research in both the basic and applied human and medical genetics; (b) to integrate both professional and public education in all areas of human genetics; and (c) to provide an arena where scientists can share their research and a platform to increase or spread knowledge and understanding of human genetics among the various professionals including healthcare professionals, health policy makers, legislators, and the public [8].

Another important organization was established in 2015. The Professional Society of Genetic Counselors in Asia (PSGCA) was created as a special interest group of the Asia Pacific Society of Human Genetics. Its vision is to be the lead organization that advances and mainstreams the genetic counseling profession in Asia and ensures individuals have access to genetic counseling services. Its mission is to promote quality genetic counseling services in Asia by enhancing practice and curricular standards, research, and continuing education [29].

Local societies were also organized in many developing countries. The Indian Academy of Medical Genetics has general body meetings at least once every calendar year. The Indonesian Society of Human Genetics organizes annual meetings every year and a scientific conference every 3 years in the Indonesian archipelago. The Medical Genetics Society of Malaysia, Genetic Counseling Society of Malaysia, Genetics Society of Malaysia, and Malaysian Society of Human Genetics are some of the active societies in the field of clinical genetics in Malaysia. In Thailand, the Thailand Society of Genetics and the Medical Genetics and Genomics Association organize meetings and academic conferences in the country all year round. Often these societies collaborate and co-organize conferences together such as the biennial conferences of the APSHG [8].

Patient support groups are also very helpful in promoting awareness of genetics and genomics. The Organization for Rare Diseases India (ORDI) is a national, nonprofit, umbrella organization that represents all the patients with rare diseases in the country. It promotes public awareness of rare diseases and collaborates with different organizations to expedite diagnosis and treatment options for patients with rare diseases [30]. In the Philippines, the

Philippine Society for Orphan Disorder (PSOD), as well as other disorder-specific patient support groups, conduct various activities, not only for patients, but also for the public. Genomics Thailand, a project organized by the Ministry of Health, Science and Technology and Education, ensures public support for success through debate and discussion through face-to-face public meetings, Websites, and social media. In Taiwan, the Taiwan Rare Disease Foundation helps in educating the public.

5 Summary/conclusion

Despite the limitations in many aspects, genetics and genomics education, training, and services in the developing world have progressed in various ways. International collaborations have been beneficial in the progress of genetics and genomics education and training. Organizations, both local and abroad as well as professional and patient support groups, also allow continuing education in genetics and genomics through various activities. Awareness of the public and education of other healthcare professionals also contributed to the development of this field in developing countries.

Acknowledgments

We gratefully acknowledge the contributions of Dr. B.R. Lakshmi, Dr. Annie Hassan, Prof. Dr. Sultana MH Faradz, Prof. Dr. Meow Keong Thong, Dr. Dineshani Hettiarachchi, Dr. Yin-Hsiu Chien, Prof. Dr. Thanyachai Sura, Dr. Ma-Am Joy Tumulak, and Dr. Vu Chi Dung.

References

[1] M.A. Campion, C. Goldgar, R.J. Hopkin, C.A. Prows, S. Dasgupta, Genomic education for the next generation of health-care providers, Genet. Med. 21 (11) (2019) 2422–2430, https://doi.org/10.1038/s41436-019-0548-4.

[2] M.K. Thong, Y. See-Toh, J. Hassan, J. Ali, Medical genetics in developing countries in the Asia-Pacific region: challenges and opportunities, Genet. Med. 20 (10) (2018) 1114–1121, https://doi.org/10.1038/s41436-018-0135-0.

[3] Geography Statistics of Asia, Retrieved from https://www.worldatlas.com/webimage/countrys/aslandst.htm. (Accessed 26 April 2021).

[4] International Data: Asia. Population Reference Bureau, Retrieved from https://www.prb.org/international/geography/asia. (Accessed 26 April 2021).

[5] World Bank Country and Lending Groups, 2021. Retrieved from https://datahelpdesk.worldbank.org/knowledgebase/articles/906519-world-bank-country-and-lending-groups. (Accessed 27 April 2021).

[6] P.A. Baird, T.W. Anderson, H.B. Newcombe, R.B. Lowry, Genetic disorders in children and young adults: a population study, Am. J. Hum. Genet. 42 (1988) 677–693.

[7] I.C. Verma, R.D. Puri, Global burden of genetic disease and the role of genetic screening, Semin. Fetal Neonatal Med. 20 (5) (2015) 354–369, https://doi.org/10.1016/j.siny.2015.07.002.

[8] E.M. Cutiongco-de la Paz, B.H. Chung, S.M.H. Faradz, M.K. Thong, C. David-Padilla, P.S. Lai, S.P. Lin, Y.H. Chen, T. Sura, M. Laurino, Training in clinical genetics and genetic counseling in Asia, Am. J. Med. Genet. C Semin. Med. Genet. 181 (2) (2019) 177–186, https://doi.org/10.1002/ajmg.c.31703. Epub 2019 Apr 29.

[9] S. Fucharoen, D.J. Weatherall, Progress toward the control and management of the thalassemias, Hematol. Oncol. Clin. North Am. 30 (2016) 359–371, https://doi.org/10.1016/j.hoc.2015.12.001.

[10] A. Christianson, C.P. Howson, B. Modell, March of Dimes Global Impact on Birth Defects: The Hidden Toll of Dying and Disabled Children, March of Dimes Birth Defects Foundation, 2006. Retrieved from: https://www.marchofdimes.org/materials/global-report-on-birth-defects-the-hidden-toll-of--d2unzZI5_VWOaLZnw6iHcx7hbpMWtWzTuIOU3DabcVY.pdf. (Accessed 29 April 2021).

[11] B.R. Korf, Genetics in medical practice, Genet. Med. 4 (6) (2002) 10S–14S, https://doi.org/10.1097/01.GIM.0000040329.94756.08.

[12] World Health Organization, Medical Genetics Services in Developing Countries: The Ethical, Legal, and Social Implications of Genetic Testing and Screening, 2006, Retrieved from https://apps.who.int/iris/bitstream/handle/10665/43288/924159344X_eng.pdf?sequence=1&isAllowed=y. (Accessed 30 April 2021).

[13] World Health Organization, Community Genetics Services: Report of a WHO Consultation on Community Genetics in Low- and Middle-Income Countries, World Health Organization, 2011. https://apps.who.int/iris/handle/10665/44532.

[14] N.D. Sirisena, V.H.W. Dissanayake, Strategies for genomic medicine education in low- and middle- income countries, Front. Genet. 10 (2019) 944, https://doi.org/10.3389/fgene.2019.00944.

[15] A. de Abrew, V.H.W. Dissanayake, B.R. Korf, Challenges in global genomics education, Appl. Trans. Genomics 3 (2014) 128–129, https://doi.org/10.1016/j.atg.2014.09.015.

[16] T.A. Manolio, M. Abramowicz, F. Al-Mulla, et al., Global implementation of genomic medicine: we are not alone, Sci. Transl. Med. 7 (290) (2015), https://doi.org/10.1126/scitranslmed.aab0194.

[17] S. Aggarwal, S.R. Phadke, Medical genetics and genomic medicine in India: current status and opportunities ahead, Mol. Genet. Genomic Med. 3 (3) (2015) 160–171, https://doi.org/10.1002/mgg3.150.

[18] J.M. Lee, M.K. Thong, Genetic counseling services and development of training programs in Malaysia, J. Genet. Couns. 22 (2013) 911–916.

[19] S.Y. Yoon, M.K. Thong, N.A. Taib, et al., Genetic counseling for patients and families with hereditary breast and ovarian cancer in a developing Asian country: an observational descriptive study, Fam. Cancer 10 (2011) 199–205, https://doi.org/10.1007/s10689-011-9,420-7.

[20] C.D. Padilla, E.C. Cutiongco-de la Paz, Genetic services and testing in the Philippines, J Community Genet. 4 (2013) 399–411, https://doi.org/10.1007/s12687-012-0102-4.

[21] N.D. Sirisena, V.H.W. Dissanayake, Genetics and genomics in Sri Lanka, Mol. Genet. Genomic Med. 7 (2019), https://doi.org/10.1002/mgg3.744, e744.

[22] K.A. Leppig, Collaborations in medical genetics: 10-year history of an ongoing Vietnamese-North American collaboration, Mol. Genet. Genomic Med. 6 (2) (2018) 129–133.

[23] K.V. Whitley, J.A. Tueller, K.S. Weber, Genomic education in the era of personal genomics: academic, professional, and public considerations, Int. J. Mol. Sci. 21 (2020) 768, https://doi.org/10.3390/ijms21030768.

[24] K.E. Ormond, M.Y. Laurino, K. Barlow-Stewart, et al., Genetic counselling globally: where are we now? Am. J. Med. Genet. 178 (2018) 98–107, https://doi.org/10.1002/ajmg.c.31607.

[25] N.J. Elackatt, Genetic counseling: a transnational perspective, J. Genet. Couns. 22 (2013) 854–857.

[26] M. Abacan, L. Alsubaie, K. Barlow-Stewart, et al., The global state of the genetic counseling profession, Eur. J. Hum. Genet. 27 (2019) 183–197, https://doi.org/10.1038/s41431-018-0252-x.

[27] M.Y. Laurino, C.D. Padilla, M.B.A. Alcausin, C.T. Silao, E.C. Cutiongco-dela Paz, A master of science in genetic counseling program in the Philippines, Acta Med. Philipp. 45 (2011) 7–9.

[28] K.Y.W. Wangi, I. Sakinah, D.A. Ningsih, N. Vanawati, D. Adiningsih, E.J. Simatupang, Nursing genetics and genomics education in Indonesia: philosophy and ethics study (a hermeneutics approach), in: The 4th International Virtual Conference on Nursing, KnE Life Sciences, 2021, pp. 857–869, https://doi.org/10.18502/kls.v6i1.8764.

[29] M.Y. Laurino, K.A. Leppig, P.J. Abad, et al., A report on ten Asia Pacific countries on current status and future directions of the genetic counseling profession: the establishment of the professional society of genetic counselors in Asia, J. Genet. Couns. 27 (1) (2018) 21–32, https://doi.org/10.1007/s10897-017-0115-6.

[30] Organization for Rare Diseases India, Vision, Mission and Objectives, 2021. https://ordindia.in/about-us/history-vision-and-mission/.

Learning outcomes, competencies, and their use in genomics education

Michelle Bishop

Learning and Training Programme, Wellcome Connecting Science, Wellcome Foundation, London, United Kingdom

1 Introduction

The discussion and debate within the literature regarding genetics, and now genomics,[a] education, and training have long moved from asking whether there is a need to educate all healthcare professionals in this area. However, we are still facing questions regarding the appropriate level of knowledge and skills required and how best to deliver this information in different educational and healthcare settings. A number of international efforts have endeavored to answer these questions through the development of competency frameworks in genetics and genomics.

I have been employed by the National Health Service (NHS) in England for over a decade, and in that time, I have worked for both the National Genetics Education and Development Centre and the Health Education England Genomics Education Programme. Both of these programs have a national focus and have been tasked with increasing the genetic and genomic competence of all the healthcare professionals who work for the NHS. While over the years we have evolved our strategic approach due to the changing priorities of the health system, there is one common feature that has prevailed throughout the years and that has been the development of competency frameworks.

In this chapter, I will share my knowledge and experience in developing and integrating these frameworks into the academic and professional training pathways. I will also provide an overview of some of the educational theories that underpin our work and define the terminology

---

[a] As some of the work that I will be talking about still contains the word "genetic" in the title, I will be using the term "genetics" where appropriate. In our program, we tended to use "genetics" up to 2009/2010 and then started to refer to "genetics and genomics." Now, we use the word "genomics." This change in terminology reflects how genetics and genomics have been integrated into the NHS.

that we use when talking about this area of education and training. Furthermore, examples of competency frameworks developed in other healthcare systems will also be described.

Before examining the "ins and outs" of competencies in genomics, the first section of this chapter will be setting the scene, providing a brief overview of the NHS and the two government-funded programs that have been established to support workforce development in genomics.

2 Setting the scene: Genetics, genomics, and the national health service in England

The National Health Service (NHS) in England is the single biggest integrated healthcare system in the world and provides care throughout an individual's lifetime, including prenatal screening, emergency care, treatments for long-term conditions, and end-of-life care. The NHS is the fifth largest employer in the world, with the NHS in England employing around 1.3 million people, of whom around 700,000 have a clinical role [1].

In 2002, The Wellcome Trust and the English Department of Health commissioned the Public Health Genetics Unit in Cambridge to develop a strategy for the delivery of genetics education and training to all health staff employed by the NHS in England, to ensure the workforce's competence in genetics [2]. Burton's (2003) report informed many of the recommendations outlined in the English Government's White Paper "Our inheritance, our future. Realising the potential of genetics in the NHS" [3]. The White Paper outlined the government's vision for the provision of genetic services in the NHS at that time. One of the highlighted areas in the White Paper was ensuring that the NHS workforce was equipped with enough genetics knowledge to allow patients and their families to benefit from genetic information and interventions now and in the future. To operationalize this vision, the National Genetics Education and Development Centre (NGEDC) was established in 2004, with the aim to help drive and coordinate genetics education for nongenetic health professionals in the NHS. In 2014, the NGEDC was formally transferred to Health Education England and evolved to become what is now known as the Genomics Education Programme [4].

Genomics became a focus within the NHS in England when the previous Prime Minister, David Cameron, announced in 2012 the ambitious aim to sequence 100,000 NHS patients' genomes. This announcement led to the establishment of the 100,000 Genomes Project, which met its goal of sequencing 100,000 whole genomes from NHS patients in December 2018 [5].

Building on the success of this project, NHS England launched a new NHS Genomic Medicine Service in October 2018, which provides, along with other genomic tests, whole-genome sequencing for selected clinical indications for rare and inherited diseases, as well as a small number of cancer diagnoses. The expectation is that the use of genomic testing will expand over time [6].

Because of the introduction of genomics into mainstream care, genomics will, to some extent, impact on many of the 1.3 million individuals working within the NHS in England, whether that is by offering a genomic test, handling the samples (including transportation), interpreting results, feeding back results to patients, or using the genomic information arising from a test to guide management and treatment. The challenge we have in the NHS is to provide appropriate education and training to all relevant healthcare professionals at scale across multiple professional groups amid this rapid pace of change.

Health Education England (HEE) is responsible for improving the quality of patient care through education, training, and development of the NHS staff in England. The Genomics Education Programme (GEP), which sits within HEE, is now the NHS's method of ensuring its staff have the knowledge, skills, and experience to ensure that the health service remains a world leader in genomic and precision medicine.

Prior to describing some of the work of the NGEDC and the GEP and the competencies and learning outcomes that have been developed and implemented in the UK, I first want to spend some time considering two educational theories that underpin this work. I then want to define competencies, and how they differ from competence, and provide an overview of the international landscape in terms of the development of learning outcomes and competencies in genetics and genomics.

3 Using theory to guide practice: The alliance between outcome-based education and adult learning theory

"*I don't need to become a mini clinical geneticist.*" This is a statement we often hear at the GEP when we connect with new groups of healthcare professionals about genomics education. Healthcare professionals want education and training that is relevant to their practice. What they do not want is to be overloaded with extraneous information that is best suited to those that specialize in this clinical area. So how do we, in the GEP, design educational interventions that ensure only relevant information is included for the healthcare professional group in question?

Many educational programs these days are designed using the broad principles of outcome-based education theory, and the courses and programs developed by the GEP are no different. Developing an education or training intervention in this way starts with considering the holistic view of what the program or course expects participants will have achieved when they have completed their course of study. That is focusing on the desired results of the education intervention, expressed in terms of the outcomes, hence the name of the theory. It is more than just the knowledge or the skills they have gained—it instead describes what the participants will be able to do with the information and skills they have learnt [7]. A vital component of this approach is considering how the outcomes reflect the complexities of the real world, and, most importantly for the work of the GEP, how the outcomes achieved by learners are relevant to their "life roles" [8].

Back in 1994, William Spady outlined four fundamental principles that underpin outcome-based education: clarity of focus, designing back, high expectations, and expanded opportunities [9]. While not all these principles can be applied to our setting in the GEP, the one about "designing back" is key to our work. This means that the development of a curriculum, or course plan, starts by first defining the desired end results. That is starting with the vision of what you want a person to be able to do when they have completed the course or program of study. Then, you work backward, by identifying all the building blocks, or the knowledge, skills, and behaviors, that learners must achieve to fulfill these outcomes. All decisions about what content is included in the course, or program of study, should be considered by asking whether the content will directly support learners in attaining the defined outcomes. If there is content that is being considered that does not support these outcomes, the course designers should be asking themselves why this content is needed. This way, only relevant information is included in the educational or training offering.

The same process can be used when creating small discrete learning opportunities for continuing professional development. In our team, when we are designing an educational or training course, we always start with the wanted end outcome—that is what we want healthcare professionals to be able to know or do in their practice after they have completed the course, or, in other words, what attributes do we want them to have? We have found that to effectively define these outcomes, and to ensure we consider the complexities of the learner's "life role" as a healthcare professional, it is good practice to involve representatives of the target audience at this point of course development. For instance, if the genomic educational intervention is for nurses, we need to involve nurses in defining the course outcomes so that we can ground the learning in the complexities of nursing practice. This ensures that the desired program outcomes are considerate of all aspects of nursing practice and the clinical pathways in which they work. Therefore, the course design strives to providing learning that can be applied to the nurse's clinical work, or their "life role" after the educational event.

Valuing learning that is relevant to real-life situations is one of the tenets of the assumptions outlined by Malcolm Knowles on how and why adults learn, which are listed in Table 1 [10]. As stated by Knowles, adults become ready to learn the things they need to be able to cope effectively with their everyday life. Another of Knowles' assumptions of adult learning is that adults' orientation to learning is problem- or task-centered, as compared to subject-centered which is the norm for school-level education. Adults also learn more effectively when the learning material is presented in the context of a real-life situation [11], such as using clinical scenarios to guide the learning activity.

By drawing on the principles of both outcome-based education and adult learning theory, our team ensures any education or training initiative that we develop is "learner-focused." Theory is only taught if it is needed to support the competent performance of tasks, not because the teacher, or educator, thinks it is interesting. In other words, there must always be a purpose for why the learners need to know this information or skill. By focusing on the outcomes and asking what a learner needs to know and do so they can achieve these outcomes, the appropriate information will be included in the educational intervention.

TABLE 1 Assumptions of adult learning theory.

1	Adults need to know why they need to learn something before they learn
2	Adults are independent and can therefore self-direct their own learning
3	They have accumulated a great deal of life experience, which is a rich resource for learning
4	They value learning that is relevant to the demands of their everyday life
5	They are more interested in immediate, problem-centered approaches than in subject-centered ones
6	They are more motivated to learn by internal drivers than external drivers

4 Defining terminology: Making sense of competence, competency, learning outcomes, and learning objectives

I find that, like the terms "genetics" and "genomics," people may conflate the terms "competence" and "competency" as well as "learning outcomes" and "learning objectives." Just as with "genetics" and "genomics," these terms do hold different meanings and appreciating these differences will support a common understanding and therefore a consistent use of terminology, by those who are involved in developing genetics and genomics education and training materials.

4.1 Competencies versus competence

Competencies, or a singular competency, describe the professional attributes, which, when performed to a certain level, result in an effective performance of a task, or set of tasks or duties. For healthcare professionals, competencies usually define the comprehensive profile of the expected knowledge, skills, and professional behaviors required for safe clinical practice.

Looking at each of these components in turn, knowledge can be considered the condition of being aware of something that has been acquired through learning, training, or experience. It can also be viewed as having the theoretically understanding of a subject area. Knowledge is often subcategorized as "descriptive knowledge," or knowledge of particular facts, and "procedural knowledge," or the knowledge of how to perform a skill or task. A skill is considered the physical ability to perform an activity. It includes physical movement, coordination, dexterity, and the application of procedural knowledge. Behaviors, which may also be referred to as "attitudes," refer to the mind-set or an individual. That is do they think and behave in a way that is required for them to effectively carry out their duties or tasks?

A competency framework is the structure that sets out the individual competencies required by individuals to perform a specific activity, area of practice, or what is expected by everyone within a professional group.

Competence, on the contrary, states that an individual can perform a specific task in a manner that yields desirable outcomes, that is to perform a competency to an agreed standard. This definition implies that to be deemed competent an individual can successfully apply the appropriate knowledge, skills, and behaviors that underpin a competency in both new and familial situations for which prescribed standards exist [12]. To assess competence, an individual's proficiency level needs to be compared to an agreed standard, measuring how someone undertakes or completes a task or competency. The person's proficiency level is then compared with this standard, where they are considered either competent in that task, or gaps in either their knowledge, skills, and/or behaviors are identified.

4.2 Learning objectives versus learning outcomes

When considering the difference between learning objectives and learning outcomes, it can be as simple as considering the words "objective," that is the purpose and intent, and "outcome," the way something, in this case the learner, turns out.

Learning objectives are, by their nature, considered from the educator's perspective, describing what the educator wants to accomplish through the education or training intervention. Despite coming from the educator's perspective, if they are planned as part of an outcome-based framework, learning objectives are still learner-centered and provide learners a sense of what knowledge or skills they will obtain by completing the course. Well-designed learning objectives can also support course developers, by steering the planning and development process, thereby ensuring that the final course or program meets the intended purpose.

Learning outcomes can be seen more from the learner's viewpoint and outline what learners should achieve by completing the course. As they provide a very clear idea of what the desired outcome looks like, providing learning outcomes upfront is considered good practice, so that learners will be able to identify what knowledge and skills they will know and be able to do after completing the course. They can then judge whether the course in question meets their learning need requirements. However, this also means that learners will have expectations about what the course will and will not teach them. Learning outcomes also help learners see how the course content and summative assessments are relevant to them.

How can learning outcomes help those that are developing and delivering the course? Focusing on the learning outcomes provides an opportunity for the course creators and educators to "put themselves in their learner's shoes," allowing them to consider whether the learner will truly be able to meet the learning outcomes at the end of the course. If they cannot, then the course content needs to be reconsidered; otherwise, the course may fail to meet the expectations of the learner. In addition, the learning outcomes can, and perhaps should, be used to guide assessment. This is also one of the tenets of outcome-based education where there is concordance between outcomes and assessment [13]. Effective learning outcomes are designed so that they provide measurable criteria that allow the educator and the learner to assess whether the course has achieved the desired outcomes.

Learning outcomes also need to reflect how the learner is anticipated to work with the knowledge provided. For example, are they expected to *recall* information such as being able to respond to the question "what is a gene?" or to *critique* the use of different genomic tests by comparing relevant information and making a judgment? Bloom's taxonomy, first published in 1956 [14], provides a structured framework for categorizing educational outcomes based on the level of complexity of the outcome's ask. The initial framework had six main categories: knowledge, comprehension, application, analysis, synthesis, and evaluation. These categories were revised in 2001 [15] to reflect the dynamic nature of learning and to signify how we, as learners, encounter and work with knowledge. Sticking with the six categories, the revised framework now includes remember, understand, apply, analyze, evaluate, create.

These categories can be viewed as lying on a continuum with the simpler outcomes on the left, starting with "remember," and the more complex tasks to the right ending with "evaluation." Identifying where on the continuum the outcomes lie requires an introspective look at what you want learners to be able to do with the knowledge provided in the course. However, I would also provide a note of caution for those that develop short courses to consider how realistic the outcome is, being mindful about the length of the course (20-min e-learning versus master's level module) and the baseline level of preexisting knowledge of your learners. Unrealistic outcomes can mean that the learning material does not meet the expectations of the learners who are completing the education or training course.

5 Genomic learning outcomes and competencies: The international landscape

There have been numerous efforts over the last twenty or more years to define the core learning outcomes and competencies for genetics, and later, genomics for healthcare professionals by key organizations throughout the world. There is universal recognition that the expected competence, and therefore underpinning competencies, will differ by health professional group. However, much of the work in this area has categorized the healthcare workforce into only two groups: the "genetic" specialist, which includes anyone whose core work revolves around genetics and genomics such as clinical geneticists or genetic counselors, and the "nongenetic" specialists, which encompasses the rest of the clinical workforce. Presented the following are a few select examples that provide a flavor of what has been achieved.

In 2001, the National Coalition for Health Professional Education in Genetics (NCHPEG), based in the United States of America (USA), developed a list of core competencies in genetics that were deemed as essential for all healthcare professionals. The objective of these competencies was to "validate the importance of a basic foundation in genetics for health care, foster the use of common terminology, increase the consistency of genetics-education efforts across the disciplines, promote active discourse about the relative role of different professionals in the provision of genetic services, and reduce duplication of effort." The original list of 44 core competencies, which were drafted and agreed upon by a multidisciplinary working group, were reviewed and revised over the years before the third edition containing 18 competencies was released in 2007. These competencies are still in circulation and can be accessed via the Jackson Laboratory website [16]. They were designed to be used by educational institutions when designing training curricula and by practicing health professionals to identify their own learning needs.

Since then, other groups in the USA have either used the NCHPEG competencies as a resource document to refine and contextualize the list for distinct professional groups [17], or have forged their own path in terms of defining competencies. One such group that has taken a different approach is the Inter-Society Coordinating Committee for Physician Education in Genomics, or the ISCC. The working group within the ISCC that was tasked with developing a competency framework for physicians recognized that the specific application of genomics will differ for different medical disciplines thereby "making it impossible to identify a single set of competencies that apply to all areas of practice" [18]. They therefore took the approach of developing a set of competencies that could be used as a starting point for anyone developing education for physicians on genomics, whereby individual competencies could be "drawn down" from the common framework to develop a customized set of competencies for a specific area of practice.

In Europe, the education committee of the European Society of Human Genetics (ESHG), as part of a collaboration with the education arm of the EuroGentest[b] project, set out to develop core competencies in genetics for health professionals working in Europe. The committee had

---

[b] EuroGentest is a project funded by the European Commission to provide consistency in the process of genetic and genomic testing across Europe. More information can be found at: http://www.eurogentest.org/index.php?id=160 (Accessed 10 July 2021).

a major undertaking as there are significant differences in how health professionals are trained and regulated throughout Europe. They therefore agreed to develop a set of core competencies that could be applied to all health professionals in Europe, regardless of their country of origin. Competencies in this context were defined as "the set of behaviors that are expected by independent professionals, which include using communication skills as well as knowledge and clinical reasoning." In contrast to the approach used in the USA, the healthcare workforce in Europe was delineated into two distinct groups: specialists in genetics (e.g., clinical geneticists) and those healthcare professionals who are generalists or who specialize in an area of health care other than genetics. Core competencies were developed for each group and are presented in distinct documents on the ESHG website [19]. It was anticipated that these competencies were to be used by educational institutions and professional societies to guide the genetic (and now genomic) education and training of healthcare professionals across Europe.

In Australia, the approach taken in developing core competencies in genetics is comparable to that taken by Europe, with an education committee established under the governance of the peak professional society, the Human Genetics Society of Australasia (HGSA). To date, the Education Committee of the HGSA has developed recommendations on the core capabilities in genetics for medical graduates [20]. A working group of genetics experts was established to define the "extent and level of genetic knowledge, skills, and behaviors that all doctors need at the start of their careers." The resulting 13 core capabilities in genetics were grouped under the headings of core knowledge, core skills, and attitudes. These core capabilities were accompanied by a set of learning objectives that support each of the capabilities.

The development of a set of generic "core" genomic competencies that could be applied in different health settings has been earmarked as an opportunity for global collaboration so that each country does not need to start this work from scratch [21]. When comparing the different competencies described earlier, there are common themes. However, each set of competencies, or capabilities, has been developed considering the nuances of the health system in question with only one program, ESHG, purposively designing a framework that could be applied by different countries. Therefore, despite these altruistic intentions of defining a set of universal competencies, the reality appears to be that each country prefers to consider the needs of their own health system and the healthcare professionals who would be delivering the clinical genomic services. Perhaps a more pragmatic approach to take the work forward would be one like that used by Tognetto et al. in Italy [22]. This group used the existing frameworks described previously as the basis to inform the development of their own country-specific genomic competencies. The resulting framework has therefore been reviewed, revised, and contextualized to the nuances of the country's health system and the differing roles of healthcare professional groups in terms of how they interact with genomic services.

6 Competencies, learning outcomes, and the continuum of genomics education in the NHS

With the introduction of whole-genome sequencing as a frontline test in the NHS in England from 2020, and the subsequent expansion of genomic services across many clinical specialities, there is now, more than ever, a focus on how the NHS will ensure all relevant healthcare staff are competent to deliver a genomic medicine service [6,23,24].

There has been a recognition for many years that there is a requirement for genetics, and now genomics, to be integrated into the academic and NHS clinical training programs[c] [3,4,24]. Although the frequency with which health professionals in the NHS encounter genomics will depend on their role and speciality, it is important that at qualification, healthcare professionals are familiar with the core genomic concepts that are needed to fulfill their professions' role in the genomics clinical pathway. Those designing preregistration training programs need to ensure graduates have the required genomic knowledge, skills, and behaviors needed for safe clinical practice in genomics at the point of registration. They also need to prepare graduates for the advances that will come from an increased understanding of the genomic contribution to common disorders and responses to medication and the changes in genomic technologies.

Education in genomics, however, should not be viewed as a static event but seen as beginning with preregistration education and continuing in professional development and workplace learning, or what has been called the "continuum of genomics education" (see Fig. 1). The genomic requirements differ at each stage of the education continuum, with knowledge gained during preregistration training laying the foundation, which can be built on during speciality training where genomic information is contextualized to the specific clinical area in which they are training. The continuum then continues after formal training to encompass lifelong learning, focusing on the competencies required to perform specific activities relevant to individual roles.

The continuum of genomics education

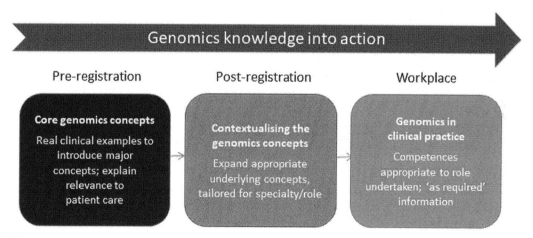

FIG. 1 The continuum of genomics education.

[c] NHS training programs: Formal qualifications with the aim of developing future NHS workforce include undergraduate or preregistration (e.g., nursing/medical school) and postgraduate training. Graduates of these programs are eligible to register with statutory regulatory bodies, such as the General Medical Council, and work within the NHS.

6.1 The continuum of genomics education: The medical professions

One of the early successes of the NGEDC was the development of agreed learning outcomes in genetics for medical students (preregistration) and medical speciality trainees in nongenetic specialities (postregistration). These learning outcomes were developed through a consensus approach, involving the itemization of the knowledge, skills, and behaviors from a range of "nonspecialists" and "experts" that formed the learning outcome working group. These statements were then refined and reviewed by the group until a consensus was reached on what the minimum genetic knowledge, skills, and behaviors required at each level of the education continuum [25,26]. These lists were then developed into learning outcome statements. The learning outcomes for medical students were endorsed by the Joint Committee on Medical Genetics and medical students teaching leads and subsequently accredited and ratified by relevant organizations. Following consultation with speciality education boards and other key stakeholders, there was widespread agreement that the learning outcomes for nongenetic medical speciality trainees were appropriate for a wide range of training programs [26]. Both sets of learning outcomes have since been revised to consider the impact of genomics on clinical care and are outlined in Tables 2 and 3.

The GEP is now working alongside the organizations that develop postgraduate medical training curricula to support the integration of these genomic learning outcomes into all relevant programs. To identify the applicable learning outcomes for each curriculum, the GEP has used an approach founded on the outcome-based education model where the GEP, alongside other stakeholders, has defined the expected genomics' attributes for graduates of different medical and surgical training programs based on what they would be expected to know and do at the completion of their training. This ensures the appropriate mix of genomic learning outcomes is included in each curriculum, so that the genomics education and training for a general practitioner will be different from that of, say a medical oncologist, reflecting their different clinical roles, the patient populations they care for, the clinical presentation of their patient groups, and ultimate use of genomic information in their practice.

One example of successful integration of genomics learning outcomes into an established curriculum is that of general practice training. General practitioner (GP) trainees, like all future healthcare professionals, want their training and any associated teaching to be relevant to their future practice [28]. Therefore, to ensure that any genomics education

TABLE 2 Learning outcomes in human genetics and genomics for medical students.

1	Understand and describe the structure and function of the human genome and how it is transmitted
2	Have an understanding of how genome variation arises and its role in health and disease
3	Be aware of how information from the genome is obtained and how it can be used to identify the genetic variation associated with disease and to inform clinical management
4	Be able to identify patients and families with, or at risk of, a condition with a strong genetic component
5	Be able to communicate genomic information in an understandable, nondirective manner, being aware of its potential impact on an individual, family, and society
6	Know how to obtain up-to-date information and scientific and clinical applications of genomics

TABLE 3   Learning outcomes in human genetics and genomics for doctors in training in nongenetic specialities [27].

1	Describe the structure, function, and transmission of the human genome, and how its expression is regulated through genetics and epigenetic mechanisms (as the basis for understanding human diseases)
2	Be able to describe and differentiate the types and relative frequencies of variations in the human genome, how they arise, and how they contribute to health and disease through their effects on gene expression
3	Know the genetic or genomic basis for disorders encountered in the speciality, and whether they follow a specific mode of inheritance
4	Recognize, in conjunction with a family history or as an isolated presentation, the subset of patients whose condition has a genetic basis (e.g., as a single gene condition) and institute appropriate genetic management for them and their family
5	Order tests using genetic technologies appropriately, interpret the results, and apply them in the context of clinical management
6	Be able to communicate genetic and genomic concepts, probability information, and test results, to facilitate engagement and choice for the patient and their family in managing their condition

initiatives align to real-life clinical scenarios, the GEP, in 2017, led a Delphi-style survey to establish the different ways in which genomics may present in primary care. Participants who responded to the survey represented a range of views including GPs, staff from clinical genetics, other practitioners working in primary care as well as the patient and public voice. The results indicated that genomic presentations include patient-led queries as well as clinical-led enquiry and actions. Participants in the Delphi survey were then asked to consider what knowledge, skills, and attitudes would be required by GPs to manage these genomic issues. The knowledge, skills, and attitudes (KSAs) identified by this Delphi survey were categorized as preexisting KSAs regarding genetics and genomics, mapping to those learning outcomes as outlined in Table 3, novel KSAs related to new technologies and the evolving clinical applications of genomic information, and finally core communication and professional KSAs [29]. The first two categories informed the content of the RCGP Curriculum Topic Guide "Genomic Medicine" [30]. The third category was acknowledged as being sufficiently covered within core GP professional and communication skills, highlighting that GPs are already well equipped with many of the KSAs required to address genomics issues presenting within primary care.

6.2 The continuum of genomics education: The nonmedical professions

The other healthcare professional group that the NGEDC initially focused on was nursing and midwifery. In fact, the NGEDC's nursing program started prior to the publication of the Government's White Paper in 2003. At the same time as Burton's seminal work was published [2], a competency-based genetics education framework for the nursing professional groups was reported [31]. Using the NCHPEG competency statements as a template for discussion, seven core competency statements were agreed to through a nominal group approach. Participants in this process represented key stakeholders from throughout the UK including managers, policy makers, and clinical educators. While there are reports that individual nursing and midwifery programs have used this competence framework to develop specific learning outcomes for their own use, there is no firm evidence that this framework has been incorporated into nursing and midwifery education at a national level [32]. There was some concern that the consultation process did not sufficiently account for the views of the healthcare professionals on the front line. As such, it was felt that this framework was based on the experts' perceptions of the educational needs of the healthcare professional groups, rather than the educational needs of the healthcare professionals themselves [33]. Kirk et al. addressed this concern by engaging front-line staff in a review of the framework to ensure it met the needs of those healthcare professionals who actively provide patient care [34]. A similar program of work was completed to develop an equivalent competency framework for midwives [35].

6.3 Using competencies to support ongoing workforce development

At the other end of the education continuum, the NGEDC developed, in conjunction with Skills for Health, a set of workforce competencies in genetics that follow the patient pathway. These workforce competencies describe in detail how and what activities should be carried out in clinical practice and the underpinning knowledge, skills, and behaviors required and have been ratified as United Kingdom National Occupation Standards (NOS) in genetics.

NOS are developed by specified standard-setting organizations, such as Skills for Health, through consultation with employers and other relevant stakeholders. NOS have a set format and specify the performance criteria (or skills) as well as the knowledge and understanding that underpin the individual NOS. National Occupational Standards are approved by UK government regulators and available for workers in a range of different sectors, including health and social care. They can be used as benchmarks for qualifications, as well as defining job descriptions, recruitment profiles, supervision, and work appraisals. All current NOS can be found in a central repository.[d] Each NOS is developed as an individual measure of competence; however, they are grouped in "suites" that identify the professional sector for which the NOS sits. According to the latest information from the central repository (2021), there are around 900 different suits that contain almost 23,000 NOS. Each standard can be used to set a benchmark for qualifications and training and to define roles and responsibilities in job descriptions. These occupational standards are also used to gauge the performance of qualified healthcare practitioners through annual appraisals, to demonstrate they have met the level specified in the NOS when undertaking specific tasks or functions within their professional role [36].

To develop these genetic NOS, the NGEDC convened a working group to identify the core genetic activities that were performed by nongenetic healthcare professionals who had a patient-facing role as part of their clinical practice. Taking the assumption that the type of genetics activities would differ by professional role, the working group had representatives

---

[d] https://www.ukstandards.org.uk/ (Accessed 7 July 2021).

TABLE 4   National Occupational Standards (NOS) for Genetics and Genomics.

NOS ID	
GTC1	Identify where genetics and genomics are relevant in your area of practice
GTC2	Identify individuals with, or at risk of, genetic conditions
GTC3	Gather multigenerational family history information
GTC4	Use multigenerational family history information to draw a pedigree
GTC5	Recognize a mode of inheritance in a family
GTC6	Assess the genetic risk associated with a condition
GTC7	Organize a test that uses genetic technologies
GTC8	Communicate genetic and genomic information to individuals, families, and other healthcare staff
GTC9	Use genomic information in clinical decision-making

from more than 20 different healthcare disciplines to consider what genetic activities were performed by which healthcare professional in the patient pathway. Using this information, the common genetic activities were identified, redefined as competencies, and the underpinning knowledge, skills, and values required to perform each competency were outlined. These nine competencies were then assessed against the quality criteria set down by the UK National Occupational Standards Panel and ratified as NOS in 2008. In 2014, each of the genetic standards was reviewed to ensure the genomic aspects of healthcare were considered and are now categorized as the National Occupational Standards for Genetic and Genomics as listed in Table 4.

Competencies can also be used to develop a flexible workforce. The needs of a healthcare system change at pace, usually in a time frame quicker than the years it takes to train clinical professionals. Therefore, the NHS is aiming to build a more flexible workforce, so that already qualified healthcare staff can learn new knowledge and develop skills over the course of their career enabling them to take on new clinical responsibilities. This provides a way to address any short-term issues in producing newly qualified clinical staff that are equipped to use new genomic technologies. The way in which the NHS is tacking this issue is by "defining sets of skill-based competencies that can apply across different professional groups" [6], that is, instead of looking at the competencies that are needed by each professional group, looking at the competencies that are required to perform a clinical activity that could be performed by a range of appropriately trained healthcare staff from different clinical backgrounds.

The GEP has initiated work in this area by developing two competency frameworks that bookend the genomic testing pathway: facilitating consent for genomic testing and the return of results [37,38]. Both of these frameworks are designed to guide best practices for any healthcare professional who may be offering germline genomic testing. Conversations about genomic testing, including the return of results, are likely to be led by individuals from different clinical backgrounds depending on the context of the test and the clinical pathway. What these frameworks provide are the agreed competencies, and therefore expected knowledge and skills, that need to be met by anyone who is performing these tasks, regardless of professional background. There is some flexibility in the adoption of these competencies; not all of the competencies outlined in the framework will be applicable to all roles or clinical pathways. Therefore, these competencies must be considered within the individual's scope of practice.

The development of these frameworks was underpinned by outcome-based educational theory. Using a consensus methodology, based on the nominal group technique used by other groups to develop competency frameworks [35], healthcare professionals from various disciplines reviewed clinical scenarios in iterative rounds to identify the key activities that would need to be performed to facilitate the consent process and to return genomic test results. The resulting activities, and underpinning knowledge and skills, were itemized and then voted on by the group as core knowledge, skill, or behavioral elements for these clinical tasks. The agreed statements were then reworked into a series of competencies. These draft frameworks were sent out to be reviewed by individual healthcare professionals, and medical royal colleges.[e] In addition, a consultation process took place with patient communities to ensure the patient narrative was included within the competencies. These frameworks have been adopted by different professional groups to guide education and training, exemplars of which are described in Boxes 1 and 2.

BOX 1 Case study: Using a competency framework to develop a training session

A gene panel test for amelogenesis imperfecta, a disorder of tooth development, was added to the NHS England Genomic Test Directory in April 2021. For pediatric dentists, accessing this test was a new, and therefore unfamiliar, part of dental care and, as such, was identified as an area requiring additional training. The Amelogenesis Imperfecta/Dentinogenesis Imperfecta Clinical Excellence Network UK (CEN) approached the GEP for guidance on available resources and worked collaboratively to develop and facilitate a training session focusing on the consent process. The facilitating genomic testing: a competency framework, described earlier in the chapter, was used to structure a virtual work-shop focusing on the competencies around communication about the genomic test, and utilized role-play scenarios to illustrate specific skills. The competency framework provided clear guidance and boundaries for the content of the workshop. The competencies also informed the questions used as part of the workshop evaluation, providing an opportunity for participants to self-assess their own competence and confidence before and after the session and to identify additional learning needs. The CEN has since led the development of local standard operating procedures (SOPs) for use in practice, aligning the SOPs to the competency framework.

---

[e] Medical royal colleges are generally responsible for setting the curriculum and assessment standards for postgraduate medical training in the UK. However, it must be noted that the accreditation of the curriculum rests with the General Medical Council.

BOX 2   Case study: Using a competency framework to support workforce development

A regional clinical genetics service in the UK was tasked with introducing BRCA1/BRCA2 into breast and gynecological cancer teams across the region. This meant that these tests would be accessed by healthcare professionals who had no previous experience in the facilitation of genomic testing. As part of this expansion of genomic services into mainstream care pathways, several specialist nurses were co-opted to complete a training package around the facilitation of genomic testing. The competency framework, "facilitating genomic testing," was used to guide the development of the training package which included an education plan and learning outcomes based on the competencies. A blended learning approach was used, with in-person training, facilitated by an experienced genetic counselor, supplemented by completion of online knowledge-based educational packages. Before and after each training session, the competency framework was used to check the nurses' ongoing knowledge and skill acquisition. The final stage of the education plan involved shadowing clinical appointments with the same genetic counselor who facilitated the in-person training. Following the completion of the education plan, a competency training and evidence form was completed for each of the nurses with a self-assessment of competence provided by the nurse and an objective review of competence completed by the genetic counselor. Areas for additional knowledge and skill development were identified at this point, with a plan devised to address any deficits. The competency training and evidence form is kept for the nurses' own records, to be used as evidence of continuing professional development, and their clinic manager was informed of the outcome. This approach has since been rolled out across the region.

7 Learning outcomes and competencies in practice: Examples from the NHS

Learning outcomes are typically used to define what has been achieved by completing a particular educational event whether these are formal academic programs or short courses. Competencies, on the contrary, can and have been used in a variety of ways. The following are some examples of how the genetic and genomic learning outcomes or competency frameworks have been used.

7.1 Integration of learning outcomes into formal academic programs

The development of the learning outcome frameworks is only one step in the process. For them to have any influence on increasing the genomic competence of the healthcare workforce, they need to be integrated into relevant academic education or training programs. The GEP has responded to part of this challenge, by influencing the key regulatory bodies that define the standards required by healthcare professionals at the point of registration. These standards define the expected attributes that a prospective healthcare professional must have in order to register with their regulatory body. If we consider the outcome-based education theory described earlier in this chapter, these standards are the literal definition of the holistic view of the expectation of what students will have achieved when they have completed their academic preregistration course of study. All the key regulators of healthcare professionals in the UK have these standards and academic institutions must comply with them and demonstrate how their syllabus meets the standards in order to be approved as an accredited provider.

At the time this chapter was written, genomics was included in the General Medical Council's (GMC) Outcomes for Graduates [39], the Nursing and Midwifery Council's (NMC) standards of proficiency for registered midwives [40], and, for the first time, the NMC standards of proficiency for registered nurses [41]. This now means that all nursing and midwifery schools, as well as medical schools, must include genomics as part of their syllabus.

The next challenge is to ensure there is consistency in the breadth and depth of genetic and genomic topics that are covered within the programs of the different academic institutions. The learning outcomes developed by the NGEDC can be used to guide syllabus development. They can also be used as a benchmark to assess the genetic and genomic content to ensure prospective healthcare professionals will have access to similar genetics and genomics education in their preregistration academic programs.

There are no such regulatory standards for postregistration medical training. However, an audit undertaken in 2013 showed that the majority of curricula did contain some genetic and/or genomic content [42]. However, the way in which different programs utilized the NGEDC framework varied. A small number had included some or all of the learning outcomes verbatim into their training programs [43]. Others appear to have used the learning outcomes more like learning objectives and used them to guide the development of their own learning outcomes that meet the specified requirements of their training program.

7.2 Using learning outcomes to design discrete educational events and assessment strategies

In our program's experience, we consider what the learning outcome, or outcomes, is for all of the educational and training materials that we produce, whether it is for a two-minute animation or a formal qualification in genomics. As stated earlier in the chapter, all our education and training resource development is underpinned by the outcome-based education theory. While the initial step involves defining the desired end result, the learning outcomes for the educational intervention are articulated through identifying the knowledge, skills, and behaviors that learners must achieve to fulfill these outcomes. Our courses, whether online or delivered face-to-face, will have these learning outcomes articulated within the course content, and the end-of-course assessments will be congruent with these learning outcomes. We use the learning outcomes to guide not only the summative assessment but also any formative assessment questions included throughout the course.

Just a sidenote on assessment terminology. Formative assessments are used to monitor student learning and to provide ongoing feedback. These assessments are considered low stakes; they provide an opportunity for students to try out their knowledge or skills in a low-risk environment, and to receive feedback on their learning. Formative assessments can help students develop and improve before the completion of a course. For educators, these assessments give "real-time" feedback about students' progression and identify areas where students may require additional instruction. Formative assessments can be educator-, peer-, or student-led (as a self-assessment); in formal courses, they may carry no grade, which can be a disadvantage as students may not see the benefit and therefore not complete the assessment. In our work, we tend to use formative assessments as a "knowledge check" within a course. It does not count toward the final mark, or grade, but provides an opportunity for learners to self-assess if they have understood the information that has just been presented. The aim of summative assessments is to evaluate the learning at the end of the course or unit of education by comparing the student's learning to an agreed standard or benchmark. Summative assessments are often "high stakes" and as such are often given a higher priority by students than formative assessments.

7.3 The multiple ways in which competencies can be used in practice

Like learning outcomes, competencies can also be used by educators to inform the development of education programs, by identifying the training needs of healthcare professional groups and as a framework to structure the learning and assessment. An example of how one of our competency frameworks has been used in this context is shown in Box 1. Depending on the type of education program, competencies may be formally assessed to compare an individual's proficiency in that competency is against an agreed standard.

Competencies can also be used by individuals to identify their own education and training needs through self-assessment. Similarly, competencies may be used as part of annual appraisal, to identify strengths and areas for development. When supporting healthcare professionals to train in a new area, a competency framework can be used to benchmark existing knowledge and skills and to identify knowledge and skill gaps. A case study, describing how a competency framework has been sued to support individual healthcare professionals to take on additional clinical responsibilities, is shown in Box 2.

From an organizational perspective, competencies can be used to assess the collective competence of a department, assess multidisciplinary skill mix, and identify knowledge and skill gaps. Competencies have been used through recruitment processes to help frame interview questions and benchmark candidates against agreed standards. They can also be used to evaluate whether a particular activity is being delivered by different departments, either in the same organization or across a healthcare system, in a consistent manner. Competency frameworks thereby provide an organization or healthcare system a mechanism to safeguard against different levels of healthcare delivery in different locations.

7.4 Competence develops over time

It should be noted that healthcare professionals acquire competence over time [44]. Benner's novice to expert theory, a construct theory originally proposed as the Dreyfus model of skill acquisition [45] and then applied and modified to nursing, is really applicable to all areas of healthcare practice. This theory espouses that an initial training opportunity, either through preregistration education or continuing professional development, leads to someone who is considered a novice, that is an individual who is new and inexperienced in the competency. After additional training and/or hands-on practical experience, that person reaches a level that, when assessed, can be considered competent. Although obtaining competence is a major milestone in professional development, it should not be considered the final point. That comes with repeated practice over time, leading to proficiency and the ultimate status of expert, which comes after many years of experience and professional growth [44]. I think this theory is summarized well by a quote from Dr. Claire Price-Dowd [46] who states:

> Competence is about being conscious of your abilities…being confident in your competence is what makes the difference in the care being delivered at the frontline.

Confidence comes with time, experience, and trusting your own capabilities.

Another one of my go-to quotes when talking about competence and competencies is the following from Secretary of State, Donald Rumsfeld:

> There are known knowns. These are things we know that we know. There are known unknowns. That is to say, there are things that we know we don't know. But there are also unknown unknowns. These are things we don't know we don't know.

What Secretary Rumsfeld is referring to here is the four stages of competence, or the conscious competence framework [47]. This model relates to the different states involved when progressing from incompetence to competence in an activity. The four stages are:

- *Unconscious incompetence*: This is where an individual does not know that they do not know something. In some instances, they may deny the usefulness of the activity. This "denial" is something that we, as educators in genomics, encounter every day and is one of the biggest hurdles that need to be overcome in the quest to ensure a genomic competent workforce. Before an individual will engage with learning, they need to recognize their own incompetence in this area, and to see value of a new clinical activity, or knowledge. This concept links in with Knowles' adult learning theory which was described earlier in the chapter.
- *Conscious incompetence*: Individuals in this stage do not have the knowledge or know how to do something, but they recognize that they have this deficit and, importantly, acknowledge the need to address this deficit. People who have identified specific learning needs fall into this category.
- *Conscious competence*: People in this stage have the knowledge and skills to "competently" perform an activity; however, performing this activity requires concentration in that there is heavy conscious involvement in the execution of the task. Individuals may break down the activity into individual steps and have checklists or standard operating procedures that they use to ensure all aspects of the activity are performed correctly. This stage can be compared to the novice/competent stage of the novice to expert continuum described earlier.
- *Unconscious competence*: Individuals who are in this stage have had so much practice in that activity that performing the task becomes "second nature." They have reached the proficiency or expert end of the "novice to expert" continuum.

8 Conclusion and summary

Competencies are used to define the set of characteristics or attributes that underlie and enable capable practice and can be used as a framework from which learning outcomes can be derived. Competencies are the preferred framework used by healthcare professional groups to assess whether practitioners are "fit for practice." These standards can be used by both educational institutes to inform the development of curricula, or by organizations to inform the development of education or training packages. They can also be used by individual health professionals as a way to identify their own learning needs.

As genomics becomes integrated into mainstream clinical care, more healthcare professional groups will have access to genomic technology to inform patient diagnoses, treatment, and clinical management. This will require a genomic literate and competent healthcare workforce, so that patients can benefit from these genomic advances. Those organizations responsible for the education and training of the different healthcare professional groups will be looking for competency frameworks and sets of learning outcomes that they can use to guide the development of education and training resources.

While there have been numerous efforts around the world to define competencies and learning outcomes in genetics and genomics, as of yet there has been no international agreed set of core competencies. Instead, different countries are taking an iterative approach, building on previous work to refine and contextualize competencies and learning outcomes to their own health setting and areas of clinical need. Although the empirical evidence of adoption of these frameworks into education and training programs is limited, the approaches taken have been evidence-based and underpinned by educational theory. This means that the competencies, and learning outcomes, reflect the health professional groups' current practice both in terms of the clinical activities carried out and the context of their clinical work or training and are therefore much more likely to be adopted in practice.

Acknowledgments

I would like to thank Amanda Pichini, Genetic Counselor at Bristol Regional Clinical Genetics Service, and the members of the Amelogenesis Imperfecta/Dentinogenesis Imperfecta Clinical Excellence Network UK for sharing their experiences in using the competency frameworks that allowed me to pull together the two case studies.

References

[1] NHS Digital, NHS Workforce Statistics—March 2021, NHS Digital, London, 2021.

[2] H. Burton, Addressing Genetics, Delivering Health. A Strategy for Advancing the Dissemination and Application Of Genetics Knowledge Throughout Our Health Professions, Public Health Genetics Unit, Cambridge, 2003.

[3] Department of Health, Our Inheritance, Our Future. Realising the Potential of Genetics in the NHS, Department of Health, London, 2003.

[4] I. Slade, D.N. Subramanian, H. Burton, Genomics education for medical professionals—the current UK landscape, Clin. Med. 16 (2016) 347–352.

[5] Genomics England, Genomics England, Department of HEalth and Social Care, 2018. December 5 https://www.genomicsengland.co.uk/the-uk-has-sequenced-100000-whole-genomes-in-the-nhs/#:~:text=Health%20Secretary%20Matt%20Hancock%20has,whole%20genomes%20from%20NHS%20patients.&text=and%20an%20automated%20analytics%20platform,genome%20analyses%20to%20th. (Accessed 10 July 2021).

[6] National Health Service, The NHS Long Term Plan, National Health Service, London, 2019.

[7]   E.S. Holmboe, R.M. Harden, Outcome-based education, in: R.M. Harden, D. Hunt, B.D. Hodges, J.A. Dent (Eds.), A Practical Guide for Medical Teachers, fifth ed., Elsevier, London, 2017, pp. 114–121.

[8]   A.-M. Morcke, T. Dornan, B. Eika, Outcome (competency) based education: an exploration of its origins, theoretical basis, and empirical evidence, Adv. Health Sci. Educ. Theory Pract. 18 (2013) 851–863.

[9]   W. Spady, Outcome-Based Education: Critical Issues and Answers, American Association of School Administrators, Arlington, VA, 1994.

[10] M.S. Knowles, The Modern Practice of Adult Education: From Pedagogy to Andragogy, The Audlt Education Company, New York: Cambridge, 1980.

[11] M.S. Knowles, E.F. Holton, R.A. Swanson, The Adult Learner: The Definitive Classic in Adult Education and Human Resource Development, Elsevier, London, 2011.

[12] D.D. Lane, V. Ross, Defining competencies and performance indicators for physicians in medical management, Am. J. Prev. Med. 14 (3) (1998) 229–236.

[13] R. Killen, Outcomes-Based Education: Principles and Possibilities, University of Newcastle, Faculty of Education, Newcastle, Australia, 2000. Unpublished manuscript.

[14] B.S. Bloom, M.D. Engelhart, E.J. Furst, W.H. Will, D.R. Krathwohl, Taxonomy of educational objectives: the classification of educational goals, in: Handbook I: Cognitive Domain, David McKay Company, New York, 1956.

[15] L.W. Anderson, D.R. Krathwohl (Eds.), A Taxonomy for Learning, Teaching, and Assessing. A Revision of Bloom's Taxonomy of Educational Objectives, Longman, New York, 2001.

[16] Genetics, National Coalition for Health Professional Education in, Core Competencies in Genetics (Online), The Jackson Laboratory, 2007. https://www.jax.org/education-and-learning/clinical-and-continuing-education/ccep-non-cancer-resources/core-competencies-for-health-care-professionals. (Accessed 5 July 2021).

[17] J. Jenkins, K. Calzone, Establishing the essential nursing competencies for genetics and Genomics, J. Nurs. Scholarsh. 39 (2007) 10–16.

[18] B.R. Korf, A.B. Berry, M. Limson, A.J. Marian, M.F. Murray, P.P. O'Rourke, E.R. Passamani, M.V. Relling, J. Tooker, G.J. Tsongalis, L.J. Rodriguez, Framework for development of physician competencies in genomic medicine: report of the competencies working group of the inter-society coordinating committee for physician education in genomics, Genet. Med. 16 (2014) 804–809.

[19] The European Society of Human Genetics, Core Competences in Genetics for Health Professionals in Europe (Online), The European Society of Human Genetics, 2010. https://www.eshg.org/index.php?id=139. (Accessed 5 July 2021).

[20] Human Genetics Society of Australasia, Core Capabilities in Genetics for Medical Graduates, Human Genetics Society of Australiasia, 2008. August 17 https://www.hgsa.org.au/documents/item/250. (Accessed 6 July 2021).

[21] T.A. Manolio, M. Abramowicz, F. Al-Mulla, W. Anderson, R. Balling, A.C. Berger, S. Bleyl, A. Chakraverti, W. Chantratita, R.L. Chisholm, V.H.W. Dissanayake, M. Dunn, V.J. Dzau, B.-G. Han, T. Hubbard, et al., Global implementation of genomic medicine: we are not alone, Sci. Transl. Med. 7 (2015) 1–8.

[22] A. Tognetto, M.B. Michelazzo, W. Ricciardi, A. Federici, S. Boccia, Core competencies in genetics for healthcare professionals: results from a literature review and a Delphi method, BMC Med. Educ. 19 (2019) 19.

[23] M. Richards, Diagnostic: Recovery and Renewal—Report of the Independent Review of Diagnostic Services for NHS England, NHS England, London, 2020.

[24] Government, HM, Genome UK: The Future of Healthcare, HM Government, London, 2020.

[25] P. Farndon, C. Bennett, Genetics Education for health professionals: strategies and outcomes from a National Initiative in the United Kingdom, J. Genet. Couns. 17 (2008) 161–169.

[26] S. Burke, M. Martyn, thimas, H., and Farndon, P., The development of core learning outcomes relevant to clinical practice: identifying priority areas for genetics education for non-genetics specialist registrars, Clin. Med. 9 (2009) 49–52.

[27] National Genetics Education and Development Centre, Learning Outcomes in Genetics and Genomics for Specialty Trainees in Non-genetic Specialties, National Genetics Education and Development Centre, Birmingham, 2014.

[28] S. Burke, M. Martyn, A. Stone, C. Bennett, H. Thomas, F. Farndon, Developing a curriculum statement based on clinical practice: genetics in primary care, Br. J. Gen. Pract. 59 (2009) 99–103.

[29] M. Bishop, J. Hayward, I. Rafi, Opportunities for education and learning in primary care genomics, InnovAiT 14 (2021) 78–84.

[30] Royal College of General Practitioners, The RCGP Curriculum: The Curriculum Topic Guides, Royal College of General Practitioners, London, 2019.

[31] M. Kirk, K. McDonald, S. Anstey, M. Longley, R. Iredale, Fit for Practice in the Genetics Era. A Competence Based Education Framework for Nurses, Midwives and Health Visitors, University of Glamorgan, Pontypridd, Wales, 2003.

[32] H. Skirton, K. Murakami, K. Tsujino, S. Kutsunugi, S. Turale, Genetics competence of midwives in the UK and Japan, Nurs. Health Sci. 12 (2010) 292–303.

[33] O. Barr, R. McConkey, Health visitors' perceived priority needs in relation to their genetics education, Nurse Educ. Today 27 (2007) 293–302.

[34] National Genetics Education and Development Centre, Fir for Practice in the Genetics'/Genomics Era: A Revised Competence Based Framework for Nurse education. Preliminary Report, National Genetics Education and Development Centre, Birmingham, 2010.

[35] E.T. Tonkin, H. Skirton, M. Kirk, The first competency based framework in genetics/genomics specifically for midwifery education and practice, Nurse Educ. Pract. 33 (2018) 133–140.

[36] R. Harrison, L. Mitchell, Using outcomes-based methodology for the education, training and assessment of competence of healthcare professionals, Med. Teach. 28 (2006) 165–170.

[37] HEE Genomics Education Programme, Facilitating Genomic Testing: A Competency Framework (Online), HEE Genomics Education Programme, 2019. https://www.genomicseducation.hee.nhs.uk/consent-a-competency-framework/. (Accessed 10 July 2021).

[38] HEE Genomics Education Programme, Communicating Germline Genomic Results: A Competency Framework (Online), HEE Genomics Education Programme, 2021. https://www.genomicseducation.hee.nhs.uk/communicating-germline-genomic-results-a-competency-framework/. (Accessed 10 July 2021).

[39] General Medical Council, Outcomes for Graduates, General Medical Council, London, 2018.

[40] Nursing and Midwifery Council, Standards of Proficiency for Midwives, Nursing and Midwifery Council, London, 2019.

[41] Nursing and Midwifery Council, Future Nurse: Standards of Proficiency for Registered Nurses, Nursing and Midwifery Council, London, 2018.

[42] National Genetics Education and Development Centre, Genomics in Heathcare education: A Review of National Curricula, National Genetics Education and Development Centre, Birmingham, 2013.

[43] Joint Royal Colleges of Physicians Training Board, Specialty Training Curriculum for Cardiology, Joint Royal Colleges of Physicians Training Board, London, 2016.

[44] P. Benner, From Novice to Expert: Excellence and Power in Clinical Nursing Practice, Commemorative Edition, Prentice Hall, Upper Saddle River, 2001.

[45] S. Dreydus, H. Dreyfus, A Five Stage Model of the Mental Activities Involved in Directed Skill Acquisition, California University Berkeley Operations Research Center, Berkeley, California, 1980.

[46] C. Price-Dowd, Confidence and competence is what makes the difference on the frontline, Br. J. Nurs. 26 (2017) 900–901.

[47] O.E. Gullander, Conscious competency: the mark of a competent instructor, Can. Train. Methods 5 (1974) 231–246.

Genomic education and training for clinical laboratory workforce

Alison Taylor-Beadling[a], Anneke Seller[b], and Simon Ramsden[c]

[a]North London Genomic Laboratory Hub, Association of Clinical Genomic Science, Great Ormond Street NHS Foundation Trust, London, United Kingdom [b]Genomics Education Programme, Health Education England, Birmingham, United Kingdom [c]Manchester Centre for Genomic Medicine, Manchester, United Kingdom

1 Introduction

If genetics refers to the study of individual genes and their role in inheritance it is tempting to think of genomics, the study of an organisms complete set of DNA, as a relatively recent discipline. In fact, genomic medicine can trace its roots back to 1959 when Lejeune and colleagues first demonstrated that children with Down syndrome had an extra copy of chromosome 21. Since which time, the discipline of medical genetics and genomics has not followed a gradualist trajectory but has been punctuated by periodic technological developments.

- In the mid 1970s, a range of chromosome banding techniques were developed allowing the recognition of individual chromosomes and the analysis of chromosomal variation.
- In the 1970s, the so-called first-generation sequencing technologies emerged, namely the Maxam-Gilbert and the Sanger (or dideoxy) methods—this latter becoming the predominant of the two approaches.
- In 1983, the polymerase chain reaction (PCR) was developed for amplifying DNA and quickly became the method of choice for purifying and cloning target sequences rapidly and facilitating more efficient mutation detection.
- In the early 1990s, comparative genomic hybridization (CGH) became available which provided an alternative means of genome-wide screening for copy number variation, offering better resolution and higher throughput than conventional cytogenetic methods.
- In the mid 2000s, the second-generation sequencing was introduced, greatly increasing the speed of DNA sequencing and facilitating a hitherto unprecedented degree of automation.

These technological advances were "born" in the academic and commercial environments but were rapidly assimilated into healthcare by the biomedical genomic workforce. Originally this workforce was split between the disciplines of cytogenetics and molecular genetics (with relevance to large genomic changes or nucleotide variation, respectively). As a result, the workforce followed one of these two different professional pathways to practice. However, since the advent of CGH, the traditional split between these two disciplines has disappeared and necessitated an alternative professional divide, reflecting a different range of skills. It is now commonplace to recognize three professional groups that make up the workforce of a modern genomics laboratory.

Technologists (also known as practitioners): Individuals with the full range of laboratory skills required for test development and optimization, troubleshooting test validation, and automation.

Bioinformaticians: Individuals who generate, analyze, and curate the huge amount of electronic data generated by modern genomic analysis.

Scientists: Individuals who are skilled in interpreting the clinical consequences of the data generated by the aforementioned two groups.

There is significant overlap between these groups including (but not exclusively) data interpretation, research and development, and quality management in all of its forms.

By recognizing both the differences and areas of overlaps, we can manage the training and regulatory requirements of these three groups and ensure appropriate workforce planning for this rapidly moving and exciting discipline.

Other workforce groups are essential to the smooth running of a genomics laboratory including laboratory assistants (booking in samples, carrying out basic laboratory functions) and administrative staff. These groups are key to a successful laboratory; however, their skills are not specific to the discipline and will not be considered further here.

Finally, it is important to note that modern genomics technologies have been adopted by many and varied routine pathology disciplines including infectious disease (microbiology and virology), transplant matching, and drug prescribing. While for the purpose of this chapter, we will concentrate on the modern genomics workforce delivering rare disease and cancer diagnostics, readers will find much in common with other disciplines.

In this chapter, we describe the underlying principles behind an effective modern genomics workforce. Our examples are drawn from practice developed in the United Kingdom over many years; however, the lessons learned will have wider relevance to practice in other countries. Table 1 provides a summary of useful resources relevant to current practice in the United Kingdom.

2 Scope of practice

Having recognized the different workforce groups within the modern genomics laboratory, we must next consider the minimum knowledge, skills, and experience as well as the main duties and responsibilities required to carry out the role. This is the scope of practice, and it is the means to developing a necessary regulatory framework for these professional groups. The scopes of practice will differ for each workforce group, but by harmonizing the different scopes of practice across laboratories, we can ensure transparent and equitable career progression that experience shows will enhance recruitment and staff retention.

TABLE 1 Useful websites.

Title	Link	Purpose
Health Education England Curriculum Library	https://curriculumlibrary.nshcs.org.uk/	Details of available training programs in England and their curricula
Clinical scientist preregistration training	https://www.nes.scot.nhs.uk/our-work/clinical-scientist-pre-registration-training/	Details of clinical scientist training in Scotland
NHS Education for Scotland Knowledge Network	Healthcare Science Trainees and Supervisors—Healthcare Science Trainees & Supervisors (scot.nhs.uk)	Resource for Healthcare Scientists Trainees and Supervisors
Health Education England Curriculum Library	https://nshcs.hee.nhs.uk/curriculum-library/	Resource of all STP, PTP, and HSST curricula for Healthcare Scientists in England
Health Education England Practitioner Training Program	https://nshcs.hee.nhs.uk/programmes/ptp/about-the-ptp-programme/	Detailing the three-year PTP program in England

Technologists: This group will focus on the "wet laboratory" work. They are skilled in the handling and processing of the full range of samples arriving in the laboratory. Dependent on skills and experience, they will develop and validate new assays drawing on skills learned from a wide range of environments, commercial, academic, and biomedical. It is essential that they learn the biological basis underpinning the tests they are delivering and in so doing they will acquire the skills required to accurately read and understand these results. Those in a senior role will have managerial responsibilities. Increasingly the emphasis is moving to high throughput sample processing and automation.

Bioinformaticians: This group has the expertise in computer software and biosciences which allows them to develop and run software pipelines for the analysis of genomic data. They are skilled in the storage and retrieval of the vast amounts of biological data generated by the modern genomics laboratory as well understanding as the biological significance of the findings.

Scientists: This group is primarily responsible for drawing a clinical interpretation from results generated within the laboratory. They will interpret the genetic and genomic results within the context of a presenting phenotype and liaise with referring healthcare professionals to arrive at a diagnosis. This clinical/laboratory liaison is a key component to the scope of practice of this group.

It is important that, depending on experience and seniority, each of these three workforce groups also participate in research, development, and translation of new techniques and assays.

Each member of staff will have a job description that is based on the scope of practice and clearly identifies the key tasks to be performed and main accountabilities of the position.

On occasion, individuals may be asked to work outside their scope of practice. This has occurred during the COVID-19 pandemic when both scientific and technical staffs have been redeployed to other areas of the healthcare system to assist with the processing, testing, and analysis of samples from patients with suspected COVID-19. If asked to work outside their scope of practice, individuals should think about their skill set and what aspects they can transfer to their new area of work from their existing role that will allow them to be effective. It is important that they continue to stick to the basic principles of practice and follow any guidance that is in place in their new work environment. Individuals should be provided with appropriate training, support, and supervision to enable them to perform the new role safely and effectively, and also be provided with opportunities to raise concerns. They should continue to work within the limits of their competence and use their professional judgment to assess what is safe and effective practice in the context in which they are working. Any concerns when working outside of their scope of practice should be raised with their line manager or a senior staff member.

3 Core training curricula in the United Kingdom

In any scientific healthcare setting, it is essential to consider the requirement of core curricula for training. This will ensure that all staff develop suitable levels of competence to perform their role within the laboratory. These standards are key to a mobile highly skilled workforce and will ensure high public confidence in the discipline. By way of example in the United Kingdom, this has involved input from multiple organizations;

Scientists and Bioinformaticians: In England, Northern Ireland, and Wales, scientist training programs (STP) are provided by the National School of Healthcare Science (NSHCS). Different STP training programs are currently available for genomics, bioinformatics, and more recently cancer genomics modalities. The STP is a three-year program that combines both workplace and academic learning culminating in award of an MSc. The STP provides trainees with an overview of the core tests delivered by these modalities and allows them to develop their professional practice skills. The curricula for STP have been developed by representatives of the genomics community, NSHCS, Higher Education Institutions (HEI), and Health Education England (HEE). The curricula are regularly reviewed to ensure they are up to date and reflect current practices within the relevant modalities.

Since 2018 in Scotland, training of preregistration scientists in genomics has been delivered by a Scottish consortium. This takes the form of a three-year postgraduate level training scheme that consists of a variety of core and rotational modules that allow the trainee to develop the skills and competencies necessary to meet the Health and Care Professions Council (HCPC) requirements. Each module aims to provide the following: knowledge and understanding, practical training, experiential learning, and assessment. Resources are shared between laboratories in the consortium using online platforms such as knowledge hub. All trainees are subject to annual review of competence progression monitoring, and on completion of the training scheme, trainees are eligible to apply for registration via the Academy of Healthcare Science (AHCS) equivalence route or Association of Clinical Science (ACS) route 2. There is currently no formal training route to registration for bioinformaticians in Scotland.

With the rapid growth of genomic testing, there has been an increase in the demand for all workforce groups. In the United Kingdom, this has put pressure on current training programs that have struggled to meet demand and has necessitated a more flexible approach to staff development. In response, many laboratories have developed in-house training schemes outside the National schemes allowing staff members to submit a portfolio for registration via ACS route 2 in molecular genetics, cytogenetics, or developing sciences or the STP equivalence route with the AHCS. Individuals training in such ways are "preregistration scientists." For those going down the AHCS equivalence route, departments must ensure that the trainees meet the competencies of the relevant STP program and satisfy the good scientific domains of practice. For those going down the ACS route 2, host departments will endeavor that trainees are rotated around various sections of the laboratory to allow them to gain sufficient experience and knowledge to meet the specified competencies in the following domains: scientific, clinical, technical, research and development, communication, problem solving, and professional accountability.

In England, the Higher Specialist Scientist Training (HSST) program offered by the NSHCS is a five-year workplace-based training program that provides clinical scientists and bioinformaticians with an opportunity to develop the skills and knowledge to allow them to apply for consultant clinical scientist posts.

Technologists: In-house training of technologists should provide technologists with sufficient experience to enable them to apply for voluntary registration with the AHCS—see the section on registration for more details. This can be achieved by rotating technologists through the various sections of the laboratory allowing them to gain a broad experience and develop professionally.

The NSHCS offers a variety of practitioner training programs (PTP) including genetics. These are a three-year undergraduate training scheme that includes work-based and academic learning. On completion of the program, individuals will be a qualified healthcare science practitioner and are eligible to apply for professional registration.

4 Qualifications, assessment, and outcomes

Entry qualifications: Career pathways in genomics in healthcare are generally designed to be progressive with entry levels for some technical posts, for example in specimen reception, requiring successful higher school examinations in a biological subject, whereas technical posts requiring more experience, to undertake PCR or sequencing usually are graduate entry positions. Clinical scientist training programs for both genomics and bioinformatics have a minimum requirement for a degree although many applicants possess Masters or PhD qualifications prior to commencing their clinical scientist training program.

Methods of training and assessment: Formal training programs usually follow a nationally agreed curriculum with an academic component delivered by a higher education institute that is examined independently, combined with a workplace-based training program that requires the student to develop a portfolio of evidence that can be used to assess competency.

In the case of clinical scientist training in England, a certificate is awarded on completion of the program, enabling registration as a clinical scientist with HCPC, dependant on a successful outcome for both the academic Masters (level 7) component and the workplace program.

Where laboratories are undertaking informal training, for example equivalence training or "on-the-job" training, or even apprenticeships, it is important to design a realistic and pragmatic training program that meets all of the requirements of the post. Even though self-directed learning is encouraged, trainees require support and mentoring at all times. It is also good practice to ensure that trainees learn from each other so if a laboratory has a number of trainees on different routes or training programs, they should come together to share knowledge and experiences.

Assessment of competency is essential to ensuring that the staff in the laboratory are safe to practice and must be undertaken objectively and thoroughly. Training officers must have the knowledge and skills to assess a particular competency and be skilled at asking open questions and drilling down to explore the student's breadth and depth of competency in a particular area. In addition, training officers should understand the difference in competency levels that are required of a newly qualified individual versus an individual who has been undertaking the role for a longer period. In recent years, the importance of appropriately qualified and experienced training officers, especially in large laboratories, has been recognized and training officers are being encouraged to pursue further study themselves, for example to obtain a PG Cert or MSc in health professional education.

There are several methods that training officers can use to assess competency. These often include direct observation of practical skills and case-based discussions as well as tests or quizzes of knowledge and understanding of procedures and processes employed within the laboratory. Together with written assessment including essays and reviews, reflective writing, and other experiential learning, the full range of competencies can be reviewed.

5 Registration to practice

Registration of laboratory staff is important to ensure high-quality patient care and public confidence in the services provided. In the UK, all clinical scientists and bioinformaticians are registered by the healthcare professions council (HCPC). The term clinical scientist is a protected title and only those on the HCPC register can use it. This allows the HCPC to ensure that those who are awarded registration meet the required standards allowing them to practice in a safe and effective manner. This protects the public and provides reassurance that an individual using a protected title such as Clinical Scientist is competent and safe to practice. It also allows the HCPC to take actions as required against those on the register who do not meet the required standards. Protection of a title is important as it gives a profession a stronger identity and credibility, prevents misrepresentation of the profession, raises standards of care, encourages investment in the profession, and raises public awareness.

In the UK, registration of genetic technologists is currently voluntary. At present, there are three routes to registration for this group with AHCS. These are as follows:

- Direct entry via an accredited PTP program.
- PTP equivalence.
- Certificate of competence for genetic technologists.

Assessment of those applying for scientist registration via either ACS route 2 or AHCS equivalence is undertaken by at least two registered scientists working at a senior level who have experience of training in the laboratory. Training is provided to assessors by ACS and AHCS part of which comprises shadowing assessments and then being paired with an experienced assessor for their initial assessments. For technologist registration, assessors are registered technologists working at a senior level. Applications from those wishing to be assessors are approved by the professional body, i.e., Association of Clinical Genomic Science (ACGS).

For clinical scientist, the HCPC sets the standards of proficiency which are required for registration.

6 Research and development

The healthcare science workforce in genomics has a key role to play in improving patient care and outcomes through contribution to research and innovation. Genomic scientists may combine academic appointments and formal research with diagnostic service delivery commitments or contribute to applied research (for example test development, evaluation, and audit) not just as a component of entry-level training but as a common theme throughout the entire career.

Opportunities for research and development are common for most genomics practitioners as diagnostic genomic laboratories are usually based in teaching hospitals with strong links to research-intensive universities. In addition, in several countries, opportunities exist to follow integrated clinical academic career pathways at different levels depending on experience and qualification including Master's level research, PhD, and postdoctoral research fellowships, thus preparing genomic scientists to undertake independent externally funded research and combine their diagnostics careers with academic appointments.

Many laboratories also now have roles for translational scientists. Their function is to bridge the gap between research and service, working closely with both groups to optimize and operationalize assays and technologies developed in the research environment so that they are robust, with the required throughput through either manual or robotic processes, and fit for purpose (analytical and clinical sensitivity and specificity) for the designated use case in the diagnostics or real-world setting.

7 Workforce planning

The contribution of genomic testing to mainstream medicine, including primary care, is rapidly increasing. In addition, new patient and care pathways are being developed which provide an opportunity for workforce remodeling against a background of multiprofessional working and the emergence of new roles. As the number of genomic tests requested per year rises within the laboratory setting, we are seeing a requirement for the understanding of genomic technologies and their application in healthcare across many pathology disciplines, in particular in hematology and histopathology. Thus, it has never been more important for the education and training of both the prospective and the current workforce in genomics to be linked to workforce planning to ensure that the laboratory services have the right number of practitioners, at the right level, with the right skills to meet service need in order to deliver timely results to inform patient treatment and/or management.

To undertake workforce planning on a national scale accurate data on the makeup of the current genomics workforce is required. Data are needed on staffing numbers and grades, their contracted hours, as well as an estimation of retirement age (either self-reported or by age). Information on staff attrition is also helpful. When combined with the expected growth rate of the service, modeling can be undertaken to inform the numbers needed to train to support the expected growth.

Even where workforce planning is effective in increasing training numbers through formal routes, there is often an urgent need within laboratories to plug gaps through alternative means as increasing workload often outpaces the availability of formally trained staff. For example, people with relevant experience such as a PhD or other technical roles can be employed in substantive posts and train on the job, overtime compiling a portfolio of evidence that can be used to demonstrate equivalence to the STP for example and thus gain registration with the HCPC as a clinical scientist.

In addition, short courses that can be relatively easily put together and delivered through universities as part of Advanced Scientific Practice, or as standalone CPD are vital for rapid upskilling of the workforce in specific areas.

Education and training need to be linked to workforce planning to ensure future delivery of services recognizing the changing relevance of genomics in healthcare.

8 Quality management

Whilst the different elements of the genomic workforce will have distinct training needs there will also be significant areas of overlap. However, one aspect of education and training is central to the entire genomics workforce, that of continual quality management. This can be broken down into four main areas: quality planning, quality assurance, quality control, and quality improvement.

While quality management is by no means unique to the healthcare environment, it will be a new concept to many joining at entry level from an academic environment. There are many excellent texts on the subject of quality management in its general sense that is outside the scope of this chapter; however, Burnett [1] is recommended reading for aspects specific to those working in Laboratory Medicine.

9 Continual professional development

On completion of training, laboratory staff will have demonstrated competence to practice within their area of expertise and with responsibilities appropriate to their professional grade. However, it is essential to recognize that this is not an end in itself and training must continue through an individual's entire career. This is the case irrespective of whether or not an individual seeks promotion to a higher grade with more responsibility. Formally, this is known as continuing professional development (CPD) and it feels particularly relevant to a discipline as rapidly moving as genomic medicine.

CPD requires the tracking and documenting of the skills, knowledge, and experience gained within the workplace. It is a record of what somebody experiences, learns, and then applies and it must be formally recorded so that it can be monitored and reviewed as sufficient and appropriate to an individual's needs.

CPD will combine different ways of learning and personal development and must be adapted to the needs of the different professional groups and grades within the workforce.

Some examples of CPD are as follows:

- **Work-based learning**: Such as reflecting on experiences at work, considering feedback from service users or being a member of a committee. Carrying out quality management processes such as audit can be particularly informative and external quality assessment schemes can provide important feedback and an opportunity to review local practice.
- **Professional activity**: Such as being involved in a professional body or giving a presentation at a conference.

- **Formal education**: For example attending courses or carrying out research.
- **Self-directed learning**: Attending seminars and reading articles or books.
- **Management training**: Formal training to develop appropriate skills.
- **Clinical:** Observing clinic appointments or contributing to multidisciplinary meetings where results are discussed with clinicians.

One of the biggest challenges can be finding time for CPD in a busy working day, but a few simple changes can make a big difference. CPD can be as simple as reflecting on discussions with colleagues or observing influential practice in others that allow one to adapt one's own methods of working. Indeed on reflection staff may find they are undertaking CPD multiple times a day without realizing it. These instances can be recorded little and often, without needing to set aside chunks of time.

There is no single format for recording CPD, although some professional bodies and employers do provide recommendations. You might choose to keep hard copies of documents in a folder or record these electronically; however, it is important to document personal learning points and reflective notes at the same time to demonstrate progression.

There needs to be a process of transparency whereby the CPD record can be reviewed by a line manager, often during annual appraisal, and often it is a requirement of professional registration that this record can be made available to external audit. It is valuable to determine a quantitative measure of this activity so that reasonable targets can be set and conformance can be measured. By way of example, the Royal College of Pathologists advocate that one hour of education normally equates to one CPD credit and 50 credits must be accrued in an annual cycle. However, it is important to be flexible in administering such a target recognizing pro rata absence from work.

10 Summary

Genomics is set to become part of routine care and is becoming increasingly relevant to a wide range of disease specialties. The genomics laboratory workforce represents a group of highly motivated individuals, agile to change, and with a diverse set of professional requirements. Different countries and healthcare models will grow their own workforce to meet this challenge; however, themes around competence, training, professional registration, and workforce planning are common to all groups.

Reference

[1]D. Burnett, A Practical Guide to ISO 15189 in Laboratory Medicine, ACB Venture Publications, 2013, ISBN: 978-0-902429-49-9.

CHAPTER

5

Genomic education and training resources for nursing

Kathleen Calzone[a] and Emma Tonkin[b]

[a]National Institutes of Health, National Cancer Institute, Center for Cancer Research, Genetics Branch, Bethesda, MD, United States [b]Faculty of Life Sciences and Education, University of South Wales, Pontypridd, Wales, United Kingdom

1 Introduction

1.1 Genomics and nursing contributions

The reduction in genomic sequencing costs from $100,000,000 in 2001 to less than $1000 in 2021 has greatly facilitated the integration of genomics into routine healthcare because these costs rival other tests or procedures already in use [1]. Consequently, evidence-based genomic clinical applications have greatly expanded the impact of genomics on nursing practice, education, and research which span the entire healthcare continuum. To appreciate the scope of genomic influences on nursing, it is helpful to walk systematically through the healthcare continuum in Fig. 1 which illustrates general nursing practice implications.

Preconception: Individuals found to have a pathogenic germline variant may inquire about the risk to future children. They may also want to learn more about and possibly consider preimplantation genetic testing in an effort to avoid transmitting a pathogenic variant to the next generation [2]. Additionally, carriers of a recessive variant may consider having their partner tested to determine whether there is any risk to future children.

Prenatal: The rapidity in which genomic discoveries are being translated into practice is perhaps best illustrated in the arena of prenatal screening. Testing of cell-free fetal DNA was quickly implemented as a component of prenatal screening. This technology is used to screen the fetus for chromosomal abnormalities, such as Down syndrome. This testing has been found to reduce the need for chorionic villus sampling or amniocentesis which confers a greater risk to the mother and the fetus [3].

Newborn screening: The purpose of newborn screening, which utilizes dried blood spots from newborn heal pricks, is to assess for evidence of conditions in which early intervention

FIG. 1   Healthcare continuum.

could reduce the morbidity and mortality of the condition, such as phenylketonuria [4]. These screens are typically not diagnostic genetic tests but screens to identify newborns in which further genetic evaluation is required. However, research is ongoing to evaluate the accuracy, feasibility, and ethics of DNA sequencing in newborns for screening purposes including the scope of results to disclose [5–7]. For instance, should pathogenic variants in high-penetrant adult-onset conditions be disclosed because of the clinical implications for the parents? [6]

Risk Assessment: Globally, genetic testing to assess disease risk has continued to expand especially, though not limited to, the areas of cancer and cardiac diseases [8–10]. As DNA sequencing costs decreased, genetic testing moved rapidly from single-gene testing informed by personal and family history rendering a differential diagnosis to panel tests that can include multiple genes [11,12]. Current research is focused on polygenetic risk scores (PRS). These scores are based on a combination of single nucleotide variants that have been found to be associated with a given disease through genome-wide association studies. The clinical validity and utility of PRS have yet to be completely established in common complex multifactorial conditions such as cancer and heart disease but evidence continues to accumulate [13].

Screening: Germline genetic testing for disease susceptibility helps to inform screening and risk reduction recommendations. But genomics can also be used as a screening tool itself. Take the example of screening for colorectal cancer using colonoscopy. This test has suboptimal uptake mostly due to the preparation for the colonoscopy despite the fact that a proportion of colon cancer can be prevented through identification and removal of polyps, a precursor to cancer [14]. Stool DNA testing, which can identify polyps and cancer emerged as a less expensive and invasive screening tool, is included in evidence-based screening guidelines [15] and has been found to be preferred by patients eligible for this test [16].

Disease Characterization: This uses genomic technologies to inform disease prognosis and/or therapeutic options. For example, in colon cancer, microsatellite instability (MSI) which are repeating segments of DNA is increasingly being routinely assessed at the time of tumor analysis in pathology. MSI can be a feature associated with deficiencies in DNA repair which may be inherited so can assist in identifying individuals who could benefit from germline genetic testing in contrast to MSI which occurs as a result of epigenetic DNA methylation [17].

Maturity-**O**nset **D**iabetes of the **Y**oung (MODY) has been found to be associated with 14 genes and counting, with the full extent of genomic underpinning of this disease yet to be firmly established [18]. The clinical MODY phenotype varies depending on the gene involved. This variation in the clinical characteristics of the disease translates into varied therapeutic options to manage MODY [19].

Individualized Therapy: Assessing a malignancy for genetic driver variants is useful to determine whether an individual is a candidate for a therapy that specifically targets that genetic variant [20]. Tumor genomic sequencing may be performed at more than one timepoint in the disease trajectory as disease progression can be due to tumor genetic heterogeneity (genetic differences within a tumor) and/or the accumulation of additional genetic changes for which other targeted therapies may be considered [21].

Pharmacogenomics represents one of the most widely applicable applications of genomics for individualized therapy. Guidelines exist to help guide medication decision-making and dosing based on genomic single nucleotide variants. PharmGKB maintains an online annotation organized by a drug with links to all international guidelines that could inform dosing or medication selection based on genotype for >130 specific drugs [22]. Some of these drugs are common medications widely prescribed such as hormonal contraceptives or ibuprofen.

Symptom management: The management of disease and/or treatment of symptoms has long been in the domain of nursing. Currently, nurse scientists are exploring the genomics of specific symptoms and symptom clusters with the goal of identifying who has the greatest risk and improved management strategies [23].

After death: As individuals develop serious illnesses, some will consider participation in research studies which often include the collection of specimens for genomic research correlatives. DNA is stable and can be stored easily for long periods. The resulting biorepositories are essential research resources. However, some research participants may have died prior to the results becoming available. Some of these results include information that if confirmed in a living relative could have clinical relevance to biological family members. Expert recommendations are to obtain a patient release at the time of sample collection indicating who would receive this information in the event of their death [24].

In summary, the healthcare continuum illustrates that regardless of the role a nurse may take, genomics can intersect. This encompasses administrators, nurse navigators [25] (nurses who help coordinate care in complex health conditions), informatic nurses who manage the electronic health record, academic and clinical educators, direct care delivery staff, and nursing scientists. Nursing as well as all other healthcare providers can expect that genomics could intersect with their role at any point along the healthcare continuum.

1.2 Genomics: Moving from specialized services into mainstream healthcare

Realizing the true benefit of genomics to healthcare quality, safety, and improved health outcomes necessitates a transition from specialist-delivered genomic care to genomics delivered at the point of care [26]. Consider the example of pharmacogenomics, which utilizes genotype to inform medication selection and dosing and identify possible inhibitors and inducers. The purpose of pharmacogenomics is to optimize therapeutic efficacy, avoid the current trial and error approach which contributes to healthcare costs, and reduce the risk of adverse drug reactions. Infectious disease is also a good exemplar for mainstreaming genomics. Antibiotics rapidly transitioned into mainstream healthcare. Infectious disease specialists, like genomics experts, are few in number and may not be readily available in all healthcare settings. Healthcare professionals quickly learned about the challenges of mainstreaming infectious diseases when providers without an adequate underpinning in the field over-rely on antibiotics contributing to the serious issue of antibiotic-resistant infections. What can we learn from this model that could translate to genomics mainstreaming? First, effective genomic mainstreaming requires adequate and ongoing education in genomics as the evidence base continues to evolve rapidly. Second, for complicated cases, the genomic expert will always still be needed [27]. Third, there remains a shortage of geneticists and genetic counselors internationally with some countries without any provider trained and serving in a genetic counselor role [28]. Taken in total, that means the general healthcare community must increase their genomic competency and practice capacity. Additionally, genetic specialists need to continue to develop novel strategies to train and support healthcare providers, especially at the point of care.

Nurses, the largest body of healthcare providers in the world, are positioned to undertake a range of activities around genomics including taking and recording family history; offering tests; supporting decision-making; returning results; and helping individuals and families understand what the results mean for them. Some may also provide genetic counseling services with adequate training and oversight. Some countries, such as China and Japan, without genetic counselors as recognized health professionals, have begun formally training nurses in this specialty [29]. Other countries, such as the United Kingdom, train and credential both nurses and individuals from other academic backgrounds such as bioscience and psychology calling them all genetic counselors [30]. The International Society of Nurses in Genetics maintains scope and standards of practice for nurses who specialize in genetics (see Table 1). Validation of competency to practice as a genetic nurse at either the general or advanced practice for countries without other credentialing mechanisms is offered through Nursing Portfolio Credentialing Commission (see Table 2).

1.3 Nursing and genomics

All nurses regardless of academic preparation, clinical role, or specialty require education and competency in genomics. However, the depth and leveling of their training and expected genomic capacity will vary depending on their academic preparation and whether they have a genetic specialist role versus another role.

TABLE 1 Genomic competency frameworks for the nongenetic specialist nursing professions.[a]

Title	Target group(s)	Level	Country	Reference	Based on
Fit for Practice in the Genetics Era: A Competence-Based Education Framework for Nurses, Midwives, and Health Visitors—Final Report. A Report for the Department of Health	Nurses, Midwives, and Health Visitors	All. Minimum requirements at the point of registration	UK	[31]	[32]
Fit for Practice in the Genetics/Genomics Era: a revised competence-based framework with Learning Outcomes and Practice Indicators: A guide for nurse education and training	Nurses	All. Minimum requirements at the point of registration	UK	[33]	[31]
Fit for Practice in the Genetics/Genomics Era: a new competency-based framework with Learning Outcomes and Practice Indicators: A guide for midwifery education and training	Midwives	All. Minimum requirements at the point of registration	UK	[34]	[31]
Essentials of Genetic and Genomics Nursing: Competencies, Curricula Guidelines, and Outcome Indicators	Nurses	All. Minimum requirements for all registered nurses	US	[35]	[31,32]
Essential Genetic and Genomic Competencies for Nurses with Graduate Degrees	Nurses	Nurses functioning at the graduate level including advanced practice registered nurses, clinical nurse leaders, nurse educators, nurse administrators, and nurse scientists	US	[36]	[35]
Competencies of genetic nursing practise in Japan: A comparison between basic and advanced levels	Nurses	All licensed nurses including genetic nurses and additional competencies for genetic specialist nurses	Japan	[37]	Original Research (Delphi study)

Continued

TABLE 1   Genomic competency frameworks for the nongenetic specialist nursing professions—cont'd

Title	Target group(s)	Level	Country	Reference	Based on
Core Competencies in Genetics for Health Professionals in Europe 03—Suggested core competencies for health professionals who are generalists or specializing in a field other than genetics	Health professionals including nurses and midwives who are generalists or specializing in a field other than genetics	Working in clinical practice	Europe	[38]	[31,32,39]
Competencies-Genetic Nurse Specialists					
Competencies of Genetic Nursing Practise Required for General and Genetic Nurses	Japanese specialty genetic nurses	Working in clinical practice	Japan	[37]	
Genetic and Genomic Nursing: Scope and Standards of Practice, 2nd Edition	Specialty genetic nurses	Working in clinical practice	International	[40]	

a Earlier versions of competencies have been included for completeness if subsequent versions have changed substantially.

TABLE 2   Genomic nursing resources.

Resource	Summary	Website
Education		
Genetics/Genomics Competency Center (G2C2)	Compilation of peer-reviewed genomic education resources for all healthcare providers	https:// genomicseducation.net/
Global Genomics Nursing Community (G3C)	Online unfolding filmed genomic case studies	https://www. genomicscases.net/en
Telling Stories: Understanding Real Life Genetics	Text and video real-life stories on the impact of genetics and genomics on the patient, family, and healthcare professional practice	http://www. tellingstories.nhs.uk/
Genomic Implementation		
Method for Introducing a New Competency: Genomics (MINC)	Genomic integration toolkit for general nursing	https://www.genome. gov/minc/toolkit
Assessment of Strategic Integration of Genomics across Nursing (ASIGN)	A maturity matrix to guide implementation efforts and benchmark progress. Can be used at institutional, local, or national levels	https://g2na. org/index.php/ g2na-publications-reports
G2NA Roadmap for Implementation of Genomics in Nursing	Essential elements to promote genomics integration globally across nursing	https://g2na. org/index.php/ g2na-publications-reports

TABLE 2   Genomic nursing resources—cont'd

Resource	Summary	Website
Genomics Nursing Research		
OMICS Nursing Science and Education Network (ONSEN)	Website designed to facilitate nursing research through data and sample sharing, common data element resources, educational resources for conducting genomic nursing research, mentorship opportunities, and listing genetic-specific pre- and postdoctoral opportunities	https:// omicsnursingnetwork. ninr.nih.gov/
Genomic Nursing Organizations		
The Association of Genetics Nurses and Counselors (AGNC)	An organization that represents genetic counselors, genetic nurses, and nonmedical patient-facing staff in the UK and Ireland who are working in clinical genetics	https://www.agnc.org. uk/
International Society of Nurses in Genetics (ISONG)	*Serves the international nursing community to foster and advocate for scientific and professional development in genomics*	https://www.isong.org/
Global Genomics Nursing Alliance (G2NA)	Promotes nursing in genomics healthcare and works to accelerate the integration of genomics across the everyday nursing practice	https://g2na.org/
Nursing Portfolio Credentialing Commission	An organization dedicated to the offering of professional advancement nursing credentialing. Currently offers genomic nursing credentials	https://nurseportfolio. org/
Transnational Alliance for Genetic Counseling	An organization of genetic counselor educators from 18 countries aimed at enhancing international communication and collaboration	https://sc.edu/study/ colleges_schools/ medicine/centers_ and_institutes_new/ transnational_alliance_ for_genetic_counseling/ index.php

2 State of nursing and genomic competency

The ability to provide high-quality patient-centered care results from being a competent practitioner able to integrate and apply knowledge, skills, professional judgment, values, and attitudes [41]. Competencies describe the essential components required to demonstrate competence in a particular role. From an education perspective, clearly articulated competencies provide the basis for establishing curricula. In the workplace, competency-based assessments can confirm an individual's ability to perform to the standard required by their role identifying any area(s) where development is needed, and provide a valid and reliable measure that can support organizations in ensuring that the knowledge and skills of their workforce meet their strategic plans and future needs. Calzone and colleagues [42] identify nursing competence in genomics as essential to ensuring patient safety and improving health outcomes, while Sharoff makes that case that holistic nursing care is not possible without the inclusion of genomics [43].

For patients to benefit from developments in genomics, competencies for all mainstream nurses working outside specialist genetic/genomic services are needed. There is a global variation of how health services are organized and delivered within and between countries. Additionally, similar variations exist as to how nurses are prepared to enter practice and their scope of practice post qualification. The result is that currently, no single framework of competencies in genomics exists for the profession. However, work is currently underway to address this gap and define global minimum competencies for nursing that can be built on to meet country and role-specific needs [44].

Historically, core competencies in genetics for all health professionals were proposed by the National Coalition of Health Professional Education in Genetics (NCHPEG) in 2001 [32], comprising 17 knowledge items, 8 skills (with an additional 9 for those involved in genetic counseling), and 10 competencies focusing on attitudes. Competencies, specifically for nurses and in some instances midwives, were subsequently developed in the United Kingdom [31], United States [35], Japan [37], and Europe [45] (Table 1). There are also genomic competencies leveled for nurses in advanced practice roles such as nurse practitioners, nurse anesthetists, and midwives [34,46]. These competencies reflect a broader scope of practice that may include prescriptive privileges and as such, much greater genomic competency expectations such as those associated with pharmacogenomics are required. At the genetic nurse specialist level, there are scope and standards of practice for genomic nurses established by the International Society of Nurses in Genetics working with the American Nurses Association [40]. At this level, these nurses may be establishing a differential genomic diagnosis, then ordering and interpreting genetic tests. Lastly, a genomic knowledge matrix was established for doctoral-level nursing scientists integrating genomics into their nursing research [47]. At this level, nurses may be designing and conducting basic science studies exploring the role genomics plays in areas such as symptoms and pain [48]. Recently, some countries, such as Brazil, began by obtaining permission to translate the United States competencies into Portuguese so they had a framework upon which to begin genomic capacity building until they could complete competencies specific to their country [35]. These frameworks address different levels of practice and broadly encompass the activities of identification (e.g., through family history taking and diagnostic testing), ongoing care and management, and prevention (e.g., health promotion messaging and identification of increased susceptibility/risk). Core to these is the need to be competent in accessing quality information and communicating accurate and clear information to patients, families, and colleagues.

With the establishment of the Global Genomics Nursing Alliance (G2NA) in 2017 [49], G2NA has continued to expand its global network with many countries expressing a need for genomic competency guidance. To address this, G2NA is undertaking a global effort to establish minimum genomic competencies expected of every nurse regardless of academic preparation, clinical role, specialty, or access to genomic technologies [44]. A review of the existing competencies for nongenetic specialist nurses reveals a common core that fits into the following nursing practice domains: Attitudes, Assessment, Identification, Referral, Education, Care, and Support.

Barriers to progressing the integration of genomics into nursing have been mapped to 5 areas [50]:

- Deficits in awareness and knowledge among educators and practitioners (including those at senior levels) result in a lack of professional engagement in genetics-genomics
- Lack of awareness at government and regulatory body levels
- Limitations in resources including time, funding, availability of appropriate educational resources, and capacity to deliver genetics-genomics education
- Lack of attention paid to the "patient voice" and
- Limited outcome evidence, compounded by the limited integration of genetics-genomics into practice.

Ten years on and we are now beginning to see movement in many of these areas as a result of the increasing number of genomic applications within healthcare and national genomics initiatives [51].

Drivers, such as the incorporation of genomics into professional standards or licensure examinations, are needed in most countries to progress genomics education. However, even when standards and position statements refer to genomics (e.g., [52]), these may not be sufficiently detailed to ensure that users understand the scope of what is needed. This is where detailed competence frameworks that make explicit learning outcomes (to guide the planning of educational sessions) and practice indicators (to measure the application of knowledge and skills) can be of value. For areas of specialty practice, the frameworks can be refined to further define expectations [53]. Even where nursing competencies in genomics currently exist, uptake and implementation have been patchy and insubstantial and are interlinked with broader complexities of integrating genomics into mainstream nursing practice.

In the following section, we summarize what is currently known about existing competence in the education, practice, and research environments.

2.1 Educational setting

In the past, nurses were trained solely in the hospital environment. Globally, nurse preparation now takes place in both academic/educational institutions (EIs) and clinical settings with a general move toward academic settings being the primary location for foundational learning. Courses combine theory with practical content delivered through simulation and clinical placements. EIs may provide both pre- and postqualification courses at a range of vocational and academic levels (e.g., diploma, associate, bachelor's, masters, or doctoral/PhD) for nurses working in both generalist and specialist areas across the lifespan. As such, genomics needs to be incorporated appropriately and meaningfully into each curriculum, educators need to be able to teach genomics at the required level and complexity for each course, and students need to have access to clinical environments in order to develop skills and make the links between theory and practice.

Studies suggest that faculties often have "limited or no educational background in genomics" [42], and nurse educators have been found to have very similar levels of foundational understanding of genomics to that of students [54,55]. Staff may recognize the importance of the subject but can lack both core knowledge and confidence in their own ability to undertake genomic tasks [56]. This may further limit their ability to incorporate the personal experience into their teaching to demonstrate authenticity and link theory with practice. To provide high-quality education, teaching staff must be able to access learning opportunities for themselves and should be supported to do so by their managers.

Research involving nursing students (both pre- and postqualification) has generally focused on assessing the understanding of core genomics principles through the Genetic Nursing Concept Inventory (GNCI) tool (see Education Implementation-Knowledge Measurement Instruments) [55,57,58] or self-reported knowledge of genetic conditions, terminology, and comfort in undertaking specified activities [59,60]. The tool was modified for use in Taiwan and Turkey, respectively, and all studies typically describe limited levels of knowledge (although the perception of knowledge did vary) and a need for further education and training [61,62].

2.2 Clinical setting

While there are nurses in specialized roles with a greater level of knowledge and skills in genomics, for example in oncology [63] and cardiology [64], available evidence suggests that generally, genomic competency in mainstream clinical practice is very similar to that of nursing students. Typically, genomics is viewed as important, but preparation and competence are limited. The systematic review by Skirton and colleagues [65] identified studies involving nurses working in a range of clinical areas and in different countries. They found a lack of studies assessing actual nursing competence with many using self-reported knowledge. In the Genetics and Genomics in Nursing Practice (GGNPS) tool [66] (see Education Implementation-Knowledge Measurement Instruments), collecting and using family history information were selected as the gauge for competence in genetics-genomics and evidence of practice integration because it does not rely on access to technology. But like other studies, these are user self-assessments. Objective assessment of competence by peers or managers is needed but will remain a challenge until there is sufficient upskilling of the workforce to a point where they are able to assess others, and equally, nursing job roles include activities that require knowledge and skills in genomics. One example of an assessment process in clinical practice is the "sickle cell and thalassemia (SCT) counseling knowledge and skills assessment record" [67] that can be used with nurses, midwives, or any other health professionals providing genetic information and/or testing relating to SCT. This process takes into account that for individuals new in post, competence may be developed over time. Other strategies being explored in areas with electronic health records (EHR) are the development of quality measures that can objectively assess nursing performance in genomics pulling specific data points from the EHR [68]. The ongoing assessment ensures that the practitioner maintains competence as part of a continuous professional development process.

Leadership is needed at all levels and areas of nursing to drive genomics integration. Direct experience or extensive knowledge in genomics is not a prerequisite for leadership. Individuals with some interest in genomics who are viewed as credible by colleagues and are able to motivate and influence others have an important role to play [69]. While there is minimal empirical evidence on competence in genomics for this group, nurses in senior management or policy positions appear to have little exposure to genetics or genomics during training or while in frontline practice are not aware of the breadth of clinical applications genomics now offers and may not even know who are the genetic/genomic experts in their area to help with any competency initiatives [70]. Without this insight, it is difficult to provide credible leadership to others. Any organization or country aiming to embed genomics within healthcare needs to consider how best to raise the knowledge and awareness of their clinical leaders. The Genomics Education Programme within Health Education England has been delivering "master classes" specifically aimed at directors of nursing, heads of midwifery, and senior nurse leaders. A US study delivered a 12-month program supporting leadership pairs comprised of nurse educator and nurse administrator dyads at 21 Magnet®-designated hospitals, with 2 additional Magnet® hospitals serving as controls. Dyads from the interventional groups received initial training followed by monthly education and peer support enabling them to undertake a variety of engagement and educational activities with colleagues at their own hospital [71]. However, the overall progress was limited despite the training and resources the study provided. The barriers to progress consisted of allocating time for additional training, obtaining additional institutional leadership support, compiling steering groups, identifying the genetic experts in their area, completing additional landscape analyses such as policy assessments, and planning awareness and educational activities. A mean of 4 months (range 1–9 months) was needed for the implementation of awareness activities followed by educational interventions, mean 7 months (range 4–11 months). Nurse administrators were essential in obtaining institutional leadership approval, providing staff release time, allocating funding and additional staff to conduct different interventions, and investing in critical infrastructure such as modifications to the EHR for features such as pedigree documentation [72]. This study is a useful illustration of the breadth of activities that need to take place locally alongside education to support the integration of genomics into nursing.

2.3 Research setting

Research provides the evidence on which clinical care is based. Typically, the involvement of nurses in research takes one of two roles [73], as a clinical research nurse undertaking a range of activities to deliver projects and clinical trials led by others, or as a nurse researcher developing and implementing one's own program of research. Although Jones refers to "new knowledge related to the progression of nursing" in the areas of practice, education, or administration as the basis for nursing research, some nurse researchers undertake "basic" (fundamental) and translational research. In these roles, the nurse may be better described as a nurse scientist. Irrespective of role, preparation, and development of competence in genomics is necessary for clinical nurses [74] and nurse researchers/scientists [75].

3 Existing pedagogy and outcomes

For anyone tasked with providing genomics education to nurses and midwife students or clinicians, we have set the scene by outlining key practical considerations for both the academic and clinical environments. These consist of Why, What, When, Where/How, and by Whom.

3.1 Why?

In the chapter introduction, an outline of tangible clinical applications of genomics supports the relevancy of genomics across all areas of practice. The case for including genomics in prequalification education and as appropriate in postqualification training as well as for practicing nurses has been made multiple times over many years. Resistance still exists particularly in the prequalification setting where arguments are made about a "full curriculum," genomics being "niche" or restricting it to science sessions. Academic and practice educators must work with colleagues to navigate these challenges to prepare nurses to deliver current evidence-based care which includes genomics.

3.1.1 Academic

In the academic setting, taking advantage of curriculum review events can provide an important opportunity to engage colleagues in conversations around the integration of genomics into the curriculum. Tasking and training one or more members of teaching staff to serve as champions can also help facilitate genomic curricular integration [76]. For countries that license nurses by exam in addition to their entry-level academic program, integration of genomic questions into those national/regional exams is a driver of curricular content as students must be sufficiently prepared by their academic institution to pass both sets of examinations.

3.1.2 Practice

In practice settings, the target for genomic education is not just clinical nurses but nondirect patient care nurses such as informatic nurses or nursing leaders. Assessing the landscape of evidence-based genomic practice applications that apply to a specific clinical setting can be used to influence making genomic nursing competency a priority in addition to guiding educational targets. To help overcome barriers, practice settings have been found to be amendable to establishing multidisciplinary steering committees to help overcome resistance and guide programing which could include more than just nurses [77]. Another strategy that can drive genomic education in practice is a mandate for continuing education units in genomics as part of relicensure. While this has been used for other topics, this has to date not been adopted in genomics.

3.2 What?

When considering the content to be taught regardless of whether this is in an academic or practice setting, the starting point should always be aligned to the competencies needed for practice that have also been discussed (see the Section State of Nursing and Genomic Competencies and Table 1). Learning outcomes and practice indicators can guide the identification of the scope and complexity of the content required which will vary depending on whether the learner is a student nurse or midwife, a qualified general or specialized practitioner, or a nurse with graduate nursing training [78]. Without a competency focus, some educators can fall into a trap of thinking that to "teach genomics" requires an emphasis on the bioscience such as DNA, genes, and chromosomes; how the concepts are related; how changes in sequence or structure can result in alterations with a variety of outcomes; common patterns of inheritance, etc. While these concepts are undoubtedly important and provide a foundational underpinning on which to build, this does not provide clinical relevance which is essential to understand what this means to nursing practice.

The issue of relevance to nursing practice requires some degree of experiential learning in both academic and practice education environments. The challenge currently is that the integration of genomics in clinical practice may vary widely based on the area and degree of integration. Therefore, robust case examples and alternates to clinical experiences such as some of the resources described in the Clinical Genomics Case Resources section of this chapter are important for mastering the genomic content and helping build clinical skills and judgment while building the foundation amenable to future genomic knowledge gain as the genomic evidence base evolves [79].

3.2.1 Academic

The educator or curriculum manager for a particular student group needs to consider on a regular basis "What is it that theses nurses or midwives need for their practice?." This can be used as the basis from which to plan the curriculum or training session(s). For prequalification students who will work in a range of clinical settings across the lifespan, the broader ethical, legal, and biopsychosocial implications of genomics for individuals and families as well as skills including family history taking and communication need to be taught alongside core bioscience concepts. For a group of qualified nurses or midwives who for example need to learn about specific genomic applications for a patient pathway, identification of the learning content may be more straightforward. However, this group may have received little or no genomics education during their original training. Therefore, students may also require foundational content to understand the specific content relevant to their clinical role. In this situation, it may be helpful to determine the level and extent of genomic knowledge before planning or delivering teaching and consider that a class could have both students with foundational genomic training and those without.

3.2.2 Practice

Like academic settings, nurses in practice may have varying levels of genomic expertise and prior training. Therefore, determining their understanding and current competency using established validated instruments is crucial (see the Section Knowledge Measurement Instruments). This assessment can provide information about competency gaps and misperceptions that can be educational targets. Additionally, baseline and repeated follow-up measures provide critical outcome information about where deficits remain that require continued education. Critical to the practice setting is appreciating how staff transition so mechanisms to assess and educate new staff need to be addressed.

3.3 When?

3.3.1 Academic

Integration of genomics into the curriculum can vary depending on the program. Standalone content makes genomics visible, but if referring to and using the content in later classes/modules and clinical experiences (on placement or in simulation) is not done, students may not understand the application to practice. Embedding genomic content throughout the curriculum for example as part of a spiral curriculum that builds on and develops learning as the course progresses and helps make genomics intrinsic to nursing practice but requires that all faculty involved have sufficient capacity to teach the genomic content. Critical to the teaching of any theoretical content is the need to reinforce on placement and in clinical simulation settings and so the timing of these opportunities should be considered as part of the curriculum design.

3.3.2 *Practice*

Education in the clinical setting can include the standard topic-specific sessions such as grand rounds. They can also consist of educational events and courses in genomics targeting just nurses or targeting a multidisciplinary audience for example by focusing on a clinical pathway or specialty. In the practice setting, critical to any genomic education endeavor is planning how to assess and train new staff as well as providing remedial education or up-skilling for existing staff given the rapid changes in genomics.

3.4 Where/how?

3.4.1 *Academic*

Genomics can be taught in a range of environments using a range of approaches, and like any subject, an educator will adjust their approach depending on class size and environment which can vary from a lecture theater of 200 + students to a small group learning with 6. Post-COVID-19, many academic programs are planning some retention of online synchronous and asynchronous teaching including the "flipped-classroom" approach where students are introduced to a learning material before a synchronous session to maximize student participation in activities and discussions. Asynchronous learning is especially attractive to students who are combining work with training as it provides flexibility for the workforce that does not require organizing specific training days or time away from their job. The African Genomic Medicine Training Initiative (AGMT) has delivered genomics education to nurses across 11 African countries through a distributed blended-learning approach combining a range of pre- and postcontact activities with a live online flipped-class contact session [80].

Experiential learning where students observe, imitate, and model is also an important component of nurse education and educators should look for opportunities where this can occur, e.g., incorporating genomics into clinical simulation scenarios. Similarly, objective structured clinical examination (OSCE) can be used to assess clinical skills. Refer to the Resources/ Overcoming Barriers Section of this chapter for additional information.

3.4.2 *Practice*

The approaches described earlier are also relevant to those in clinical practice. Additionally, many in-person professional conferences were forced to transition to virtual due to COVID-19. An unanticipated benefit of this has been an increase in access to events as there are no travel expenses and time away from the workplace is reduced. As conferences transition to a post-COVID-19 world, many are adopting a hybrid model where both in-person and virtual options are available. Both conference and practice-specific education can be interprofessional in nature [81] as all healthcare providers need to achieve and maintain genomic competency. Additional strategies can include the addition of a genetic healthcare provider in clinical case conferences which not only provides a mechanism to integrate evidence-based genomic applications into care but also infuses clinically relevant genomic knowledge over time to the attendees which often include nurses. Otherwise, including genomic content in continuing education content such as grand rounds and other practice-specific education offerings. Lastly, in areas where considerable deficits are identified, providing genomic education series that can include didactic content, supporting clinical case reviews and journal clubs.

3.5 Whom?

3.5.1 Academic

Individuals with a range of backgrounds can be used to teach genomics to nurses. The evidence shows that nursing faculty knowledge in genomics is low and not higher than the students that they teach [54,82]. Targeting nursing faculty to increase their capacity to integrate genomics into their education is vital in both academic and clinical settings. Until nursing educators have an adequate underpinning in genomics, bringing in specialists (e.g., with a biomedical science or clinical genomics background) may be an option for some educational institutions to help overcome this deficit. However, such an approach can have limitations: it does not build genomics education capacity within its own staff; risks the topic being perceived as niche, particularly if content taught is not then reflected or reinforced in additional areas of the nursing curriculum and by other staff; scientific staff teaching genomics may not be able to relate/link the content to the clinical nursing context, thus perpetuating a theory-practice gap. Coteaching classes with biomedical and/or genomic specialists are one strategy that can both help educate the faculty and assure that the scientific context is balanced with the nursing level and clinical applications of that information [83]. Using online teaching platforms provides an opportunity to involve genomic nursing experts who are not local to the educational institution and/or patient/service users who might otherwise have difficulty traveling to the campus. Involving patients, carers, and family members in education can enrich the student learning environment and provide insight into and experience of clinical situations that would otherwise not be available to the students.

3.5.2 Practice

In the practice setting, an assessment of available genomic expertise is critical. These can include genetic counselors, geneticists, oncologists, and pharmacists as well as others. In the Magnet® Hospital project described in the Clinical Setting section, none of the dyad members in that study knew who their local genomic experts were. Therefore, an assessment of the setting for experts who could provide some of the education is important and speaks to the value of why the dyads established interprofessional steering groups [70].

3.6 Educational outcomes

Currently, there remains a lack of evidence of what constitutes effective genomics education in nursing and for other professions within the health workforce. There is a critical need for more longitudinal studies to measure knowledge outcomes including retention, changes in practice as well as whether the nurse has sufficient capacity to apply genomics in practice. The RISE2 Genomics framework will help guide education outcome reporting and enable comparison across studies that can inform best practice [84].

4 Pathways to nursing genomic competency

What follows is a selection of tools and resources that can be used to guide, support, and measure the development of genomic competence and the integration of genomics into nursing practice and education (see also Table 2).

4.1 Education implementation

Any intervention to educate nurses in genomics requires validated instruments to measure baseline knowledge to identify deficits and misperceptions to inform the education and methods to be used. Additionally, reassessment post-intervention to evaluate what level of knowledge gain was achieved and assess over time the sustainability of knowledge gain.

4.2 Knowledge measurement instruments

4.2.1 Genomics nursing concept inventory

The Genomics Nursing Concept Inventory (GNCI) is a tool designed to measure the understanding of concepts essential for nursing practice [85]. GNCI contains 31 items across 4 topics, human genome basics, inheritance, mutations, and genomics healthcare, and is based on the US Essential Nursing Competencies [35]. It has been used as a pre- and post-teaching assessment tool for US baccalaureate nursing students [86] to assess educator (faculty) knowledge [54] as a pre- and post-interventional measure for an initiative to improve faculty competence in genomics [87] and to measure the knowledge of nurses and midwives in practice [88].

4.2.2 Genetics and genomics in nursing practice survey

The Genetics and Genomics in Nursing Practice Survey (GGNPS) was built on the foundation of a validated instrument that assesses factors influencing genetic and genomic competency in family physicians [89]. That instrument measures domains of Rogers Diffusion of Innovations (DOI) [90] GGNPS is a modified version of the original instrument in the context of nursing practice. GGNPS includes eight sections that cover the following domains: attitudes, receptivity, confidence, social system, and adoption. Instrument question types include select all that apply pick lists, multiple-choice, yes/no, true/false, and Likert scales. GGNPS can be administered online or in a paper format. Items from each domain are analyzed individually and are not combined to form a total score. Instead, items alone or in combination are intended to be used to inform competency and implementation efforts [70]. Twelve questions have been used to generate a knowledge score by first grading responses as correct or incorrect and then calculating the number of correct responses out of 12. Knowledge scores can only be calculated on surveys in which the respondent answered all 12 questions. GGNPS is designed to measure practice competency and therefore use had been in clinical practice settings such as in Magnet® Hospitals [72]. The instrument has also been translated into other languages for use globally (i.e., Japanese, Arabic, and Turkish) and is open access so can be modified for the local context.

4.3 Genomic nursing organizations

There are two international nursing organizations devoted to genomics and nurses. Individual countries may also have country-specific genomic nursing organizations, such as those in Brazil, Taiwan, Japan, and the United Kingdom, but the two organizations listed below are unique in that they are international in nature.

4.3.1 *International society of nurses in genetics*

The International Society of Nurses in Genetics (ISONG) is a membership organization targeted toward genomic healthcare with support aimed at nurses with either a clinical or research focus in genomics. ISONG has developed International Scope and Standards of Practice available from https://www.nursingworld.org/nurses-books/. Additionally, ISONG hosts disease- and topic-specific special interest groups such as cancer genetics, presents educational webinars every 1–2 months, and holds an annual congress for networking and educational purposes.

4.3.2 *Global genomics nursing alliance*

The Global Genomics Nursing Alliance (G2NA) is a network of nurses working on integrating genomics into nursing practice and education [91]. G2NA has developed the Nursing Roadmap for Genomics and the Assessment of Strategic Integration of Genomics across nursing instruments [44]. The organization hosts international webinars quarterly and periodic meetings to help further genomic integration into general nursing and midwifery practice.

4.4 Educational resources

Nursing educators are challenged to integrate genomic content into academic or continuing education in part because their genomic competency may also be limited [92,93]. This is in contrast to other areas of healthcare such as body substance precautions. Therefore, educators can benefit from both a clear framework about what their discipline needs to know and do, with supporting educational tools to facilitate in providing that education (Table 2).

4.4.1 *Genetics/genomics competency center*

The Genetics/Genomics Competency Center (G2C2) is an online multidisciplinary peer-reviewed resource repository. The resource includes access to genomic competencies for all healthcare providers, such as nurses, physicians, and pharmacists. Resources undergo peer review by advisory board members to validate that the resource is current and accurate. Resources can be identified by searching a topic or searching resources that support a discipline-specific competency [94]. G2C2 was recently expanded to also include listings of resources supported by individual professional organizations.

4.4.2 Clinical genomic case resources

Nursing education requires not just cognitive and sociocultural learning approaches but also clinical experiences to achieve the knowledge and skills necessary for practice [95]. However, genomics presents a challenge given that there is an uneven clinical application of evidence-based genomic applications limiting the access to and availability of those encounters. Therefore, novel resources that can provide real-life genomic patient stories and/or clinical encounters have been developed to address this need.

4.4.3 Telling stories, understanding real life genetics

Telling Stories, Understanding Real Life Genetics is a free-to-access, award-winning website that uses verbatim narratives ("stories") from individuals with or at risk of a genetic condition, family members, and health professionals. There is evidence that patient narratives or clinical experiences shared by nurse educators can help link theory to practice and stimulate critical thinking and active engagement [96]. This resource was developed to support educators with limited experience of genetics who had no or few stories to tell and provide an alternative way for students to learn from patients when access to suitable placements during training is limited.

More than 100 stories are available as text (to read online or download in a printable format) and some have video clips that can be downloaded and incorporated into teaching presentations. Being set within an educational framework, stories have been curated by themes enabling users to find content that, for example, fits a lesson plan or topic of interest. Each story has its own toolbox of resources including points for reflection and discussion, activities, and expert commentaries to further support teaching and learning. Although originally developed for nursing, the website is also suitable for use by many other health professional groups, and "professional role" and "clinical specialism" can be used to identify relevant content. In addition, every story has a *"How does this story relate to professional practice?"* section where the team explicitly identify where the story illustrates a specific UK nursing or midwifery genomic competency [33,34].

4.4.4 Global genetics and genomics community

The Global Genetics and Genomics Community (G3C) is an online, unfolding, simulation case study website. G3C online clinical experiences consist of filmed professional actors simulating real clinical encounters. There is evidence that these kinds of simulations can be utilized effectively in place of traditional clinical experiences [97]. Cases are designed so the learner asks questions of the patient and views the filmed patient response, which leads to additional questions until the learner is prepared to make healthcare recommendations. Segments of the encounter include additional learning activities and links to evidence-based guidelines. The learner can collect medical records from the patient as applicable. Cases are driven by questions the learner asks the patient. The patient is a professional actor who has been filmed responding to each question the learner selects. Filmed responses lead to more questions from which the learner chooses. Based on the series of learner questions and patient answers, the case scenario unfolds. Throughout the encounter, additional evidence-based resources and learning activities are offered to assist in learning more about content introduced in the encounter. Medical records can be viewed as part of the case and new information such as outside test reports can be collected as part of the encounter, viewed by the learner and placed in a virtual medical record. Learning outcomes are assessed through a series of questions associated with healthcare recommendations. Following patient encounter, the learner views a commentary by experts in that field discussing the case content. All cases undergo external peer review for content accuracy prior to filming as well as review of supporting materials including learner activities and faculty/educator guides.

4.5 Genomic education and nursing practice implementation resources

Genomic implementation is complex and necessitates consideration of the structure, process, and outcomes as described by Donabedian [98]. For each of these elements, methods and corresponding resources on how to guide genomic implementation are essential to support change, as well as mechanisms to measure the outcomes.

4.5.1 Nursing roadmap for genomics

A roadmap to drive and coordinate genomics integration across all areas of nursing was established by the Global Genomics Nursing Alliance (G2NA) [99]. Combining evidence obtained through G2NA activities with implementation science principles, the practical action plan should be adaptable to local health systems and clinical and educational contexts. Sequential stages of implementation (planning, engaging, executing, and reflecting and evaluating) are exemplified through specific actions and activities that can be undertaken and supported through a set of questions that can be used to guide strategic planning.

4.5.2 Implementation assessment: Assessment of strategic integration of genomics across nursing (ASIGN)

Assessment of Strategic Integration of Genomics across Nursing (ASIGN) [44] is a self-assessment tool developed by G2NA that can be used to both benchmark the current situation of an organization and plan a strategic course for improvement. ASIGN forms an integral part of the G2NA integration roadmap (above). By completing the tool over two or more occasions, users can identify continuity and changes in their organization over time. The tool comprises 6 critical success factors (CSF) that should be in place to effectively integrate genomics into nursing practice: enhanced education and workforce development; effective nursing practice; sustainable infrastructure and resources; collaboration and communication; public and patient involvement; and healthcare transformed through policy and leadership. CSFs are divided into subthemes, also referred to as key enablers, and each is defined by a number of outcome-focused indicators. Using a 5-point scale, users identify the current circumstance for their organization for each indicator and can evidence this with locally relevant measures. In terms of education and training, ASIGN captures the need for a culture that supports capacity building for educators and learners alike. Teaching staff should be supported to develop their knowledge and skills around genomics and the content that is taught should be assessed in nursing examinations. Genomic competencies should be embedded in prequalifying/student curricula as well as integrated into all levels of continuing professional development (CPD) courses across all specialisms.

4.5.3 *Method to integrate a new competency*

The Method to Integrate a New Competency (MINC) website was developed based on the outcomes from a 1-year study of 23 US Magnet Recognition Program® hospitals (intervention and control) participating in a 12-month genomic nursing competency integration into practice research project [70]. The website includes an arm for administrators and a separate arm for educators. The website consists of a detailed implementation guide and recommendations for how to overcome barriers. Additionally, a resource repository of items developed as part of this study is included with details about how to access each item [100].

4.6 Nursing research resources

Fundamental to effective nursing practice in genomics is the generation of new knowledge that could influence quality nursing care. The complexity of this undertaking requires both training and ongoing mentorship as well as mechanisms to facilitate collaboration. To answer questions faster, robust data sharing and maximizing the use of samples are essential.

4.6.1 *OMICS nursing science and education network*

The OMICS Nursing Science and Education Network (ONSEN) website was developed to help achieve the research priorities outlined in the Genomic Nursing Science Blueprint [101]. ONSEN education resources include the genomic knowledge matrix to support the preparation of nursing scientists using or going to use genomics in their research [47]. Other education resources include a directory of individuals willing to mentor novice genomic nursing scientists and a location to post or find current genomic pre/postdoctoral opportunities. ONSEN's Common Data Element section includes a one-stop location for common data element resources relevant to nursing research. The Resource Collaboration section is intended to enhance collaborative opportunities between investigators. In addition, this section provides information on available opportunities for sharing datasets as well as samples for additional analysis and a means to identify collaborators [102].

5 Overcoming barriers

5.1 Theoretical frameworks

Multiple change theories exist that can help guide pathways forward for genomic competency and integration efforts. They can be used in isolation to focus on a specific aspect of that effort. However, it is always wise to be cognizant of other factors that may accelerate or impede activities. As such, theories can be used in combination, particularly when they contain overlapping components, to provide a more holistic view of the approach needed [103]. Individual and collective behaviors are important components of theories most commonly applied to nursing and genomics (discussed later). We suggest that reviewing psychological theory can provide additional insights into the complexities of overcoming barriers to competency and integration while informing implementation strategies.

5.1.1 Theory of planned behavior

Ajzen's Theory of Planned Behavior [104] proposes that an individual's behavior can be predicted by his/her intention to carry out that behavior, with three constructs in combination, predicting intention. "Attitude" or expected value of the behavior describes how advantages or otherwise the person views the behavior and its outcomes. "Subjective norms" encompass their perception of social pressures and the need to comply with others. Motivation to comply may vary depending on where or who the social pressure is coming from. Thirdly, it is the level of "perceived behavioral control" or self-efficacy that they have over the behavior. It should be recognized that the relative importance of each construct can vary between individuals and groups and therefore a "one-size-fits-all" approach to behavior change is unlikely to be successful across the piece.

5.1.2 Rogers' diffusion of innovations

Rogers' Diffusion of Innovations theory has been used and studied at both individual and organizational levels [90]. The theory provides a framework for constructing pathways for the adoption of genomics into nursing education and practice. Dimensions of the theory include attributes of the innovation such as complexity and perceived need, dissemination communication mechanisms, and the social system in which the innovation is being introduced for example a healthcare institution, and whether the institutional leadership values genomics including prioritizing and enabling integration efforts. Factors associated with the diffusion process include knowledge of genomics, persuasion which considers advantages, compatibility with existing practice, complexity of implementation, ease of use, and whether outcomes are observable. The theory acknowledges that the decision to adopt can change over time, so issues of sustainability are essential. Adoption is on a progressive scale from early adopters who are described as innovators to laggards who are the slowest to adopt [90].

5.1.3 The transtheoretical model (stages of change)

The Transtheoretical Model provides a framework for the process and influencing factors involved in change [105]. While this model has largely been used in the context of health behavioral changes, this has been used to assess faculty readiness to integrate genomic content in nursing education [106]. The model provides an important insight into the stages of change and what kind of interventions could facilitate movement to the next stage. Five stages have been described. **Precontemplation**-no intention or motivation to change. The value of change may not be appreciated or is considered too complex, both of which are amenable to education. **Contemplation**-recognize the value and importance, thinking about change but no action is undertaken. Access to the implementation roadmap [99] may help guide their thinking. **Preparation**-planning for implementation of the change is underway. In this stage, access to genomic education resources is important. **Action**-implementation is underway which can be time and work-intensive, such as implementing genomic course changes. And lastly, **Maintenance**-the changes are sustained for six or more months. In the rapidly evolving field of genomics, continual updating of educational content is required.

5.1.4 Opinion leaders and influencers

Diffusion of innovation takes time and is influenced in part by the possible advantages, how simple the innovation is to implement, and observability, all key issues influencing genomic adoption [90]. However, there is evidence that opinion leaders and champions may help accelerate the adoption process [103,107–110]. Opinion leaders are trusted members of a given group, such as a healthcare institution, and have enough social and communication skills to exert influence over others. Critical to considering the use of opinion leaders is that those individuals must be champions of the innovation [110]. In the context of genomics, to be both a champion and an opinion leader, individuals need adequate support and education to recognize the potential of what genomics can bring to patient care to help move genomic implementation forward. In nursing, research in both academic educational settings and healthcare institutions found that provision of training, support, and use of genomic champion opinion leaders can be a facilitator. But those studies also found adoption still takes considerable time in part because genomics is complex [70,109]. This is not surprising as most nurses and nursing faculty have little to no scientific underpinning in genomics from which to build upon so the education deficits cannot easily be addressed by simple continuing educational interventions. This is further complicated because many of the changes brought about by the evidence-based use of genomics are not observable, and this lack of observability slows rates of adoption [90]. For example, preventing disease for those with an inherited susceptibility due to a germline pathogenic variant or use of preemptive pharmacogenomic testing resulting in the right drug selection for a given patient at the right dose to achieve both expected efficacy and avoid adverse drug reactions. When something is not observable, nurses must have a sufficient underpinning in the field to do what is required in practice. Additionally, the genomic evidence base is changing rapidly requiring a sustained implementation effort. This may necessitate funding for infrastructures such as novel education tools as students receiving didactic training in genomics may still not see the use of genomics in practice in their clinical placements. In practice, the needed infrastructure may include the development of point-of-care decision support for healthcare providers or the mechanism to easily document a family history in the electronic health record. Lastly, the access to genomic experts varies widely with many settings having no access to people with this expertise.

Considering these complexities, an opinion leader given adequate support and genomic training can be valuable to championing genomic competency and integration efforts [70,109]. Furthermore, healthcare institution nurse champions were found to engage other healthcare professionals to establish multidisciplinary genomic implementation teams to help further institution-wide genomic competency and implementation initiatives [70,111].

5.2 Critical role of leadership

As discussed earlier, any kind of academic or institutional genomic competency and implementation effort requires leadership support. Financial support for training and other implementation efforts is essential. Protected time and release time from other duties for champion opinion leaders and/or faculty and staff to receive training and have the capacity and support to move forward with implementation efforts are critical. Depending on the setting, this may require upper-level approval such as the Dean or an institutional board of directors [70,109]. But those approvals alone appear to be insufficient.

For instance, in academic settings, the curriculum committee or equivalent body, not the Dean is responsible for instituting changes in curricular content. In healthcare institutions, the line leader, such as a head nurse of a unit, can drive whether a change initiative is or is not sufficiently supported. Therefore, capacity building in genomics needs to focus on ALL levels of leadership [70]. Many training and implementation initiatives have focused on the end users, faculty or nurses, and other healthcare providers at the point of care. But focusing on the leadership at all levels can provide the needed support and infrastructure to accelerate genomic implementation and assure sustainability.

There are some efforts that can result in rapid leadership attention to genomic competency. Changes in licensure or registration requirements for practice are accompanied by education standards that drive curricula and course development. Academically, integrating genomics into accreditation standards for the academic institutions was found to make an immediate and rapid response on the part of academic leadership at all levels [106]. In healthcare institutions, accreditation standards for the practice settings such as the Joint Commission in the United States could also drive changes that are associated with quality and safety of healthcare which applies to genomics. However, these leaders may also be lacking in genomic capacity sufficiently enough to make changes to existing regulations. In addition, these leaders may be far harder to reach and engage, let alone have them understand the relevance genomics to their role. But there have been efforts that have pushed genomic competency and implementation forward at all levels. This has included both public demand but more importantly government lead initiatives such as the All-of-Us effort in the United States [98] and Genome UK in the United Kingdom [112] which has led to senior nurse leaders across the country placed at the highest levels of decision-making.

Reaching into these high levels of healthcare regulation can be facilitated by the engagement of scientific advisors who can serve as champion opinion leaders. But just engaging those individuals to be champions is often insufficient. The messaging of the scientific advisors must be clear, consistent, and meaningful, and there is nothing more meaningful to all levels of healthcare and health provider training than healthcare quality and safety. Genomics epitomizes healthcare quality, safety, and cost containment. Consider the case of cardiovascular disease, a common health condition globally. Preemptive pharmacogenomic testing was found to be substantially less expensive than reactive testing performed as a result of efficacy or safety indications [incremental cost-effectiveness ratio (ICER) $86,227/quality-adjusted life years (QALY)] for preemptive testing in contrast to reactive pharmacogenomic testing ICER $148,726/QALY [113].

6 Conclusion

Knowledge, skills, and behaviors/attitudes are fundamental to achieving competence which is necessary for delivering genomic healthcare, and the integration of evidence-based genomics into nursing practice impacts healthcare quality and safety [114]. However, creating a nursing workforce that has the necessary knowledge, skills, and behaviors/attitudes in genomics is a complex process that goes beyond understanding the "what, when, and how" of providing clinically relevant genomics education. Implementation capacity is influenced by the need for infrastructure, service redesign, ongoing genomic education, and leadership support. Once a vision for genomics healthcare within a health system is established, it is then possible to define the nursing requirements and set a path for delivering the end goal.

Achieving that end goal is not as straightforward as it sounds. Genomics is a complex competency with novel terminology, and many of the outcomes associated with the appropriate use of genomics, such as pharmacogenomics, are not outwardly observable. Adding to the challenge are the rapid advances in genomic technology at ever-decreasing costs, resulting in an expanding array of evidence-based genomic clinical applications. Yet, healthcare is always evolving and there have been and continue to be healthcare challenges that result in major practice changes. These include HIV and the implementation of body substance precautions and more recently the highly contagious airborne virus, SARS-CoV-2, requiring robust infection control measures. In both of those instances, nursing as well as other health professions were able to successfully pivot and make the necessary changes even though when implementing those precautions, the outcomes were just unobservable. So, why has genomics been so challenging? There is one fundamental difference. Every healthcare professional has an educational underpinning in infection control. However, in the context of genomics, especially in nursing, there is no genomic educational foundation upon which to build, and the faculty (academic or clinical) may not be sufficiently knowledgeable to teach this content [54,92].

Regardless of your education or practice setting, this chapter set out to provide a foundation in genomic clinical applications that impact nursing across the healthcare continuum. The state of genomic nursing competency, the defined nursing genomic competencies, strategies for education, educational resources, instruments to measure genomic knowledge and implementation as well as the roadmap guide your path forward. Lastly, approaches to conquer competency and implementation barriers are offered. These tools position you to overcome the complexity to achieve your own genomic competency and/or that of your constituency whomever they may be. Unlike other areas of healthcare, genomics is rapidly evolving, so it will be critical that any plan includes circling back, updating your constituency, and training new members when they join the team.

Acknowledgments

This chapter received in-kind support from the National Cancer Institute, Center for Cancer Research, Genetics Branch, USA, and the University of South Wales, UK.

References

[1] National Human Genome Research Institute, DNA Sequencing Costs: Data, 2020, 8/25/2020 [cited 2020 9/28/2020]; Available from: https://www.genome.gov/about-genomics/fact-sheets/DNA-Sequencing-Costs-Data.

[2] T.A. Rich, et al., Comparison of attitudes regarding preimplantation genetic diagnosis among patients with hereditary cancer syndromes, Fam. Cancer 13 (2) (2014) 291–299.

[3] Screening for fetal chromosomal abnormalities: ACOG practice bulletin summary, number 226, Obstet. Gynecol. 136 (4) (2020) 859–867.

[4] J. Klein, Newborn screening from an international perspective—different countries, different approaches, Clin. Biochem. 44 (7) (2011) 471–472.

[5] O. Ceyhan-Birsoy, et al., Interpretation of genomic sequencing results in healthy and ill newborns: results from the BabySeq project, Am. J. Hum. Genet. 104 (1) (2019) 76–93.

[6] L.F. Ross, E.W. Clayton, Ethical issues in newborn sequencing research: the case study of BabySeq, Pediatrics 144 (6) (2019), e20191031.

[7]   J.M. Friedman, et al., Genomic newborn screening: public health policy considerations and recommendations, BMC Med. Genet. 10 (1) (2017) 9.

[8]   T.W. Hokken, et al., Precision medicine in interventional cardiology, Interv. Cardiol. 15 (2020), e03.

[9]   H. Hampel, et al., A practice guideline from the American College of Medical Genetics and Genomics and the National Society of Genetic Counselors: referral indications for cancer predisposition assessment, Genet. Med. 17 (1) (2015) 70–87.

[10] M.T. Bashford, et al., Addendum: a practice guideline from the American College of Medical Genetics and Genomics and the National Society of Genetic Counselors: referral indications for cancer predisposition assessment, Genet. Med. 21 (12) (2019) 2844.

[11] S.N. Hart, et al., Mutation prevalence tables for hereditary cancer derived from multigene panel testing, Hum. Mutat. 41 (8) (2020) e1–e6.

[12] H. Daoud, et al., Genetic diagnostic testing for inherited cardiomyopathies: considerations for offering multigene tests in a health care setting, J. Mol. Diagn. 21 (3) (2019) 437–448.

[13] C.M. Lewis, E. Vassos, Polygenic risk scores: from research tools to clinical instruments, Genome Med. 12 (1) (2020) 44.

[14] R. Cardoso, et al., Colonoscopy and sigmoidoscopy use among the average-risk population for colorectal cancer: a systematic review and trend analysis, Cancer Prev. Res. (Phila.) 12 (9) (2019) 617–630.

[15] D.A. Ahlquist, Multi-target stool DNA test: a new high bar for noninvasive screening, Dig. Dis. Sci. 60 (3) (2015) 623–633.

[16] X. Zhu, et al., National survey of patient factors associated with colorectal cancer screening preferences, Cancer Prev. Res. (Phila.) 14 (2021) 603–614.

[17] A. de la Chapelle, Colon cancer germline genetics: the unbelievable year 1993 and thereafter, Cancer Res. 76 (14) (2016) 4025–4027.

[18] R. Naylor, A. Knight Johnson, D. del Gaudio, Maturity-onset diabetes of the young overview, in: GeneReviews®, 2018. [Internet]. [cited 2020 11/5/2020]; Available from: https://www.ncbi.nlm.nih.gov/books/NBK500456/.

[19] S.C. Oliveira, et al., Maturity-onset diabetes of the young: from a molecular basis perspective toward the clinical phenotype and proper management, Endocrinol. Diabetes Nutr. 67 (2) (2020) 137–147.

[20] N.A. Seebacher, et al., Clinical development of targeted and immune based anti-cancer therapies, J. Exp. Clin. Cancer Res. 38 (1) (2019) 156.

[21] A. Rübben, A. Araujo, Cancer heterogeneity: converting a limitation into a source of biologic information, J. Transl. Med. 15 (1) (2017) 190.

[22] PharmGKB®, PharmGKB, 2020, [cited 2020 11/6/2020]; Available from https://www.pharmgkb.org/guidelineAnnotations.

[23] M.K. McCall, et al., Symptom science: omics supports common biological underpinnings across symptoms, Biol. Res. Nurs. 20 (2) (2018) 183–191.

[24] S.M. Wolf, et al., Returning a research participant's genomic results to relatives: analysis and recommendations, J. Law Med. Ethics 43 (3) (2015) 440–463.

[25] A.L. Byrne, et al., Exploring the nurse navigator role: a thematic analysis, J. Nurs. Manag. 28 (4) (2020) 814–821.

[26] E.D. Green, et al., Strategic vision for improving human health at the forefront of genomics, Nature 586 (7831) (2020) 683–692.

[27] H. Burton, C. Jackson, I. Abubakar, The impact of genomics on public health practice, Br. Med. Bull. 112 (1) (2014) 37–46.

[28] N. Dragojlovic, et al., The composition and capacity of the clinical genetics workforce in high-income countries: a scoping review, Genet. Med. 22 (9) (2020) 1437–1449.

[29] H. Skirton, N. Arimori, M. Aoki, A historical comparison of the development of specialist genetic nursing in the United Kingdom and Japan, Nurs. Health Sci. 8 (4) (2006) 231–236.

[30] M. Abacan, et al., The global state of the genetic counseling profession, Eur. J. Hum. Genet. 27 (2) (2019) 183–197.

[31] K.M. Kirk, et al., Fit for Practice in the Genetics Era: a Competence Based Education Framework for Nurses, Midwives and Health Visitors—Final Report. A Report for the Department of Health, University of Glamorgan, UK, 2003.

[32] J. Jenkins, et al., Recommendations of core competencies in genetics essential for all health professionals, Genet. Med. 3 (2) (2001) 155–159.

[33] M. Kirk, E. Tonkin, H. Skirton, An iterative consensus-building approach to revising a genetics/genomics competency framework for nurse education in the UK, J. Adv. Nurs. 70 (2) (2014) 405–420.

[34] E.T. Tonkin, H. Skirton, M. Kirk, The first competency based framework in genetics/genomics specifically for midwifery education and practice, Nurse Educ. Pract. 33 (2018) 133–140.

[35] Consensus Panel on Genetic/Genomic Nursing Competencies, Essentials of Genetic and Genomic Nursing: Competencies, Curricula Guidelines, and Outcome Indicators, second ed., American Nurses Association, Silver Spring, MD, 2009.

[36] K.E. Greco, S. Tinley, D. Seibert, Development of the essential genetic and genomic competencies for nurses with graduate degrees, Annu. Rev. Nurs. Res. 29 (2011) 173–190.

[37] N. Arimori, et al., Competencies of genetic nursing practise in Japan: a comparison between basic and advanced levels, Jpn. J. Nurs. Sci. 4 (1) (2007) 45–55.

[38] H. Skirton, et al., Genetic education and the challenge of genomic medicine: development of core competences to support preparation of health professionals in Europe, Eur. J. Hum. Genet. 18 (9) (2010) 972–977.

[39] K. Calzone, et al., Establishing essential nursing competencies and curricula guidelines for genetics and genomics, Oncol. Nurs. Forum 33 (2) (2006) 419.

[40] International Society for Nurses in Genetics, A.N.A., Genetics/Genomics Nursing: Scope and Standards of Practice, second Edition, American Nurses Association, 2016, p. 140.

[41] M. Fukada, Nursing competency: definition, structure and development, Yonago Acta Med. 61 (1) (2018) 1–7.

[42] K.A. Calzone, et al., Introducing a new competency into nursing practice, J. Nurs. Regul. 5 (1) (2014) 40–47.

[43] L. Sharoff, The emerging genetic-genomic era and the implications for the holistic nurse, J. Holist. Nurs. 35 (1) (2017) 5–6.

[44] E. Tonkin, et al., A maturity matrix for nurse leaders to facilitate and benchmark progress in genomic healthcare policy, infrastructure, education, and delivery, J. Nurs. Scholarsh. 52 (5) (2020) 583–592.

[45] A. Tognetto, et al., Core competencies in genetics for healthcare professionals: results from a literature review and a Delphi method, BMC Med. Educ. 19 (1) (2019) 19.

[46] K.E. Greco, S. Tinley, D. Seibert, Essential Genetic and Genomic Competencies for Nurses with Graduate Degrees, 2012, [cited 2012 5/22/2012]; Available from http://www.nursingworld.org/MainMenuCategories/EthicsStandards/Genetics-1/Essential-Genetic-and-Genomic-Competencies-for-Nurses-With-Graduate-Degrees.pdf.

[47] M. Regan, et al., Establishing the genomic knowledge matrix for nursing science, J. Nurs. Scholarsh. 51 (1) (2019) 50–57.

[48] K.T. Hickey, et al., Precision health: advancing symptom and self-management science, Nurs. Outlook 67 (4) (2019) 462–475.

[49] K.A. Calzone, et al., The global landscape of nursing and genomics, J. Nurs. Scholarsh. 50 (2018) 249–256.

[50] M. Kirk, et al., Genetics-genomics competencies and nursing regulation, J. Nurs. Scholarsh. 43 (2) (2011) 107–116.

[51] Z. Stark, et al., Integrating genomics into healthcare: a global responsibility, Am. J. Hum. Genet. 104 (1) (2019) 13–20.

[52] (ACN), A.C.o.N, in: A.C.o.N. (ACN) (Ed.), Nurses, Genomics and Clinical Practice, ACN, Canberra, 2020.

[53] England, P.H, NHS Sickle Cell and Thalassaemia (SCT) Screening Programme (2020) Sickle Cell and Thalassaemia Counselling Knowledge and Skills, 2020.

[54] C.Y. Read, L.D. Ward, Faculty performance on the genomic nursing concept inventory, J. Nurs. Scholarsh. 48 (1) (2016) 5–13.

[55] S. Dewell, et al., Assessing knowledge of genomic concepts among Canadian nursing students and faculty, Int. J. Nurs. Educ. Scholarsh. 17 (1) (2020) 20200058.

[56] M.K. Donnelly, et al., Nurse faculty knowledge of and confidence in teaching genetics/genomics: implications for faculty development, Nurse Educ. 42 (2) (2017) 100–104.

[57] L.D. Ward, J. Purath, C. Barbosa-Leiker, Assessment of genomic literacy among baccalaureate nursing students in the United States: a feasibility study, Nurse Educ. 41 (6) (2016) 313–318.

[58] R. Kronk, A. Colbert, E. Lengetti, Assessment of a competency-based undergraduate course on genetics and genomics, Nurse Educ. 43 (4) (2018) 201–205.

[59] A. Maradiegue, et al., Knowledge, perceptions, and attitudes of advanced practice nursing students regarding medical genetics, J. Am. Acad. Nurse Pract. 17 (11) (2005) 472–479.

[60] C.H. Dodson, L.P. Lewallen, Nursing students' perceived knowledge and attitude towards genetics, Nurse Educ. Today 31 (4) (2011) 333–339.

[61] C.-Y. Hsiao, et al., Taiwanese nursing students' perceived knowledge and clinical comfort with genetics, J. Nurs. Scholarsh. 43 (2) (2011) 125–132.

[62] B. Kıray Vural, et al., Nursing students' self-reported knowledge of genetics and genetic education, Public Health Genomics 12 (4) (2009) 225–232.

[63] N. Percival, et al., The integration of BRCA testing into oncology clinics, Br. J. Nurs. 25 (12) (2016) 690–694.

[64] M. Kirk, E. Tonkin, M. Longley, M. Llewellyn, A. Simpson, D. Cohen, A. Edwards, Evaluation of the British Heart Foundation Cardiac Genetics Nurses Service Development Initiative, 2012.

[65] H. Skirton, A. O'Connor, A. Humphreys, Nurses' competence in genetics: a mixed method systematic review, J. Adv. Nurs. 68 (11) (2012) 2387–2398.

[66] K.A. Calzone, et al., Survey of nursing integration of genomics into nursing practice, J. Nurs. Scholarsh. 44 (4) (2012) 428–436.

[67] England, P.H, Sickle Cell and Thalassaemia (SCT) Counselling Knowledge and Skills Assessment Record, 2020.

[68] R. Kleinpell, A.N. Kapu, Quality measures for nurse practitioner practice evaluation, J. Am. Assoc. Nurse Pract. 29 (8) (2017) 446–451.

[69] V. Andrews, et al., Identifying the characteristics of nurse opinion leaders to aid the integration of genetics in nursing practice, J. Adv. Nurs. 70 (11) (2014) 2598–2611.

[70] K. Calzone, J. Jenkins, S. Culp, L. Badzek, Hospital nursing leadership led interventions increased genomic awareness and educational intent in magnet® settings, Nurs. Outlook 66 (2018) 244–253.

[71] J. Jenkins, et al., Methods of genomic competency integration in practice, J. Nurs. Scholarsh. 47 (3) (2015) 200–210.

[72] K.A. Calzone, et al., Hospital nursing leadership-led interventions increased genomic awareness and educational intent in magnet settings, Nurs. Outlook 66 (3) (2018) 244–253.

[73] H.C. Jones, Clinical research nurse or nurse researcher? Nurs. Times 111 (19) (2015) 12–14.

[74] S.M. Sharif, et al., Enhancing inclusion of diverse populations in genomics: a competence framework, J. Genet. Couns. 29 (2) (2020) 282–292.

[75] E.S. Barbato, B.J. Daly, R.J. Darrah, Educating nursing scientists: integrating genetics and genomics into PhD curricula, J. Prof. Nurs. 35 (2) (2019) 89–92.

[76] J. Jenkins, K.A. Calzone, Genomics nursing faculty champion initiative, Nurse Educ. 39 (1) (2014) 8–13.

[77] K.A. Calzone, et al., Hospital nursing leadership-led interventions increased genomic awareness and educational intent in magnet settings, Nurs. Outlook 66 (3) (2018) 244–253.

[78] K.A. Calzone, et al., Establishing the outcome indicators for the essential nursing competencies and curricula guidelines for genetics and genomics, J. Prof. Nurs. 27 (3) (2011) 179–191.

[79] S.A. Lisko, V. O'Dell, Integration of theory and practice: experiential learning theory and nursing education, Nurs. Educ. Perspect. 31 (2) (2010) 106–108.

[80] V. Nembaware, et al., The African genomic medicine training initiative (AGMT): showcasing a community and framework driven genomic medicine training for nurses in Africa, Front. Genet. 10 (2019) 1209.

[81] C.K. Rubanovich, et al., Physician preparedness for big genomic data: a review of genomic medicine education initiatives in the United States, Hum. Mol. Genet. 27 (R2) (2018) R250–r258.

[82] C.Y. Read, L.D. Ward, Misconceptions about genomics among nursing faculty and students, Nurse Educ. 43 (4) (2018) 196–200.

[83] T. Williams, R. Dale, A partnership approach to genetic and genomic graduate nursing curriculum: report of a new course's impact on student confidence, J. Nurs. Educ. 55 (10) (2016) 574–578.

[84] A. Nisselle, et al., Ensuring best practice in genomics education and evaluation: reporting item standards for education and its evaluation in genomics (RISE2 Genomics), Genet. Med. 23 (2021) 1356–1365.

[85] L.D. Ward, M. Haberman, C. Barbosa-Leiker, Development and psychometric evaluation of the genomic nursing concept inventory, J. Nurs. Educ. 53 (9) (2014) 511–518.

[86] L.D. Ward, C. Barbosa-Leiker, B.F. French, Item and structure evaluation of the genomic nursing concept inventory, J. Nurs. Meas. 26 (1) (2018) 163–175.

[87] L.M. Bashore, et al., Facilitating faculty competency to integrate genomics into nursing curriculum within a private US University, Nurs. Res. Rev. 8 (2018) 9–14.

[88] H. Wright, et al., Genomic literacy of registered nurses and midwives in Australia: a cross-sectional survey, J. Nurs. Scholarsh. 51 (1) (2019) 40–49.

[89] J. Jenkins, S. Woolford, N. Stevens, N. Kahn, C.M. McBride, The adoption of genomic-related innovations by family physicians, Case Stud. Bus. Ind. Gov. Stat. 3 (2010) 70–78.

[90] E. Rogers, Diffusion of Innovations, fifth ed., The Free Press, New York, 2003.

[91] K.A. Calzone, et al., The global landscape of nursing and genomics, J. Nurs. Scholarsh. 50 (3) (2018) 249–256.

[92] C.Y. Read, L.D. Ward, Misconceptions about genomics among nursing faculty and students, Nurse Educ. 43 (4) (2018) 196–200.

[93] C.Y. Read, L.D. Ward, Faculty performance on the genomic nursing concept inventory, J. Nurs. Scholarsh. 48 (1) (2016) 5–13.

[94] K.A. Calzone, et al., Establishment of the genetic/genomic competency center for education, J. Nurs. Scholarsh. 43 (4) (2011) 351–358.

[95] M. Stoffels, et al., How do undergraduate nursing students learn in the hospital setting? A scoping review of conceptualisations, operationalisations and learning activities, BMJ Open 9 (12) (2019), e029397.

[96] M. Kirk, et al., Storytellers as partners in developing a genetics education resource for health professionals, Nurse Educ. Today 33 (5) (2013) 518–524.

[97] J.K. Hayden, R.A. Smiley, M. Alexander, S. Kardong-Edgren, P.R. Jeffries, The NCSBN national simulation study: a longitudinal, randomized, controlled study replacing clinical hours with simulation in prelicensure nursing education, J. Nurs. Regul. 5 (2) (2014) S3–S64.

[98] A. Donabedian, Evaluating the quality of medical care. 1966, Milbank Q. 83 (4) (2005) 691–729.

[99] E. Tonkin, et al., A roadmap for global acceleration of genomics integration across nursing, J. Nurs. Scholarsh. 52 (3) (2020) 329–338.

[100] J. Jenkins, et al., Methods of genomic competency integration in practice, J. Nurs. Scholarsh. 47 (3) (2015) 200–210.

[101] Genomic Nursing State of the Science Advisory Panel, K.A. Calzone, J. Jenkins, A.D. Bakos, A.K. Cashion, N. Donaldson, W.G. Feero, S. Feetham, P.A. Grady, A.S. Hinshaw, A.R. Knebel, N. Robinson, M.E. Ropka, D. Seibert, K.R. Stevens, L.A. Tully, J.A. Webb, A blueprint for genomic nursing science, J. Nurs. Scholarsh. 45 (1) (2013) 96–104.

[102] L.A. Tully, K.A. Calzone, A.K. Cashion, Establishing the Omics Nursing Science & Education Network, J. Nurs. Scholarsh. 52 (2) (2020) 192–200.

[103] V. Leach, et al., A strategy for implementing genomics into nursing practice informed by three behaviour change theories, Int. J. Nurs. Pract. 22 (3) (2016) 307–315.

[104] I. Ajzen, The theory of planned behavior, Organ. Behav. Hum. Decis. Process. 50 (2) (1991) 179–211.

[105] J.O. Prochaska, W.F. Velicer, The transtheoretical model of health behavior change, Am. J. Health Promot. 12 (1) (1997) 38–48.

[106] J.F. Jenkins, K.A. Calzone, Are nursing faculty ready to integrate genomic content into curricula? Nurse Educ. 37 (1) (2012) 25–29.

[107] A. Reyneke, C. Jaye, T. Stokes, Local clinical pathways: from 'good ideas' to 'practicality' for general practitioners, J. Prim. Health Care 10 (3) (2018) 215–223.

[108] T.W. Valente, P. Pumpuang, Identifying opinion leaders to promote behavior change, Health Educ. Behav. 34 (6) (2007) 881–896.

[109] J. Jenkins, K.A. Calzone, Genomics nursing faculty champion initiative, Nurse Educ. 39 (1) (2014) 8–13.

[110] G. Doumit, et al., Local opinion leaders: effects on professional practice and health care outcomes, Cochrane Database Syst. Rev. 1 (2007), Cd000125.

[111] J. Jenkins, K.A. Calzone, S. Caskey, S. Culp, M. Weiner, L. Badzek, Methods of genomic competency integration in practice, J. Nurs. Scholarsh. 47 (2015) 200–210.

[112] UK Government, Genome UK: The Future of Healthcare, 2020.

[113] Y. Zhu, et al., A model-based cost-effectiveness analysis of pharmacogenomic panel testing in cardiovascular disease management: preemptive, reactive, or none? Genet. Med. 23 (2020) 461–470.

[114] A.R. Starkweather, et al., Strengthen federal and local policies to advance precision health implementation and nurses' impact on healthcare quality and safety, Nurs. Outlook 66 (4) (2018) 401–406.

Genomic education based on a shared space for discovery: Lessons from science communication

Jonathan Roberts[a,b] *and Anna Middleton*[a,c]

[a]Society and Ethics Research Group, Wellcome Connecting Science, Cambridge, United Kingdom [b]East of England Genomic Medicine Service, Addenbrookes Hospital, Cambridge, United Kingdom [c]Faculty of Education, University of Cambridge, Cambridge, United Kingdom

1 Introduction

We begin by summarizing the difference between the "dialogue" model of science communication and the "deficit" model. Understanding these helps provide the context for a more in-depth discussion about how the field of science communication can help genomics professionals.

Next, we will briefly sketch some key elements of the history of science communication. Communicating scientific information cannot be separated from context, be it cultural, political, or social. This historical perspective also helps highlight how science communication involves the creation and negotiation of boundaries: truth, authority, politics, and power.

Following this historical account, this chapter will explore how science works not just within culture, but as a *form of culture.* This will be looked at primarily through the work of Bourdieu and the concept of "science capital." This is an influential concept in science communication that has gained traction with both science institutions and governments.

Finally, this chapter will end with a discussion of the concept of "hybrid space" and how this can provide a framework for genomics professionals to think about how to engage nonexperts.

2 Deficits and dialogues

Two models of science communication that have long stood in opposition to each other are the deficit and dialogue models of science communication. Science communication is a diffuse and varied activity. In general, a deficit model assumes a hierarchy, a top-down imposition of meaning. Science is defined and created by experts and is communicated to nonexperts or lay audiences. Successful science communication in this context would be measured by knowledge. If a nonscientist's knowledge improves, science communication is deemed successful. The dialogue model stands in contrast to this. The key idea of dialogue models, as the name suggests, is that communication goes two ways. The public cannot be seen as empty vessels, waiting to be filled with scientific knowledge. What counts as successful science communication within a dialogue model is a complex and much debated question. Broadly speaking, it fosters communication between people with different perspectives (Fig. 1).

In recent years, science communication has seen a shift toward the language of dialogues [1]. We will examine this development in more detail in the next section. However, scholarship has shown that while there may have been a rhetorical shift, science communication—in many contexts—has found it very difficult to operate outside of deficit models. This is understandable when we think about what science is and how it functions in society. Science derives a great deal of its authority from the belief that it has a privileged epistemology. That is, science and scientists can make unique claims about what is true. The criteria used to demarcate science and nonscience have been subject to sustained philosophical, historical, and sociological debate. On one end of the spectrum, there are those who believe science is different from nonscience because science is value-free, objective knowledge. This might be considered similar to what Chalmers [2] described as a "common sense" view of science, where facts about the world can be established through the testing of theories by observation. Here, there is a belief in a unique "scientific method" that provides objective and value-free truths about the natural world. For our purposes here, we label this the *positivist* view, with the recognition that this is a simplification of a range of more nuanced stances.

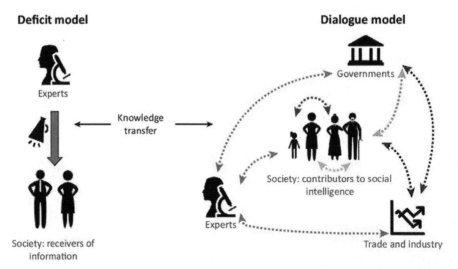

FIG. 1   Deficit and dialogue models of science communication. *Taken from F. Courchamp, A. Fournier, C. Bellard, C. Bertelsmeier, E. Bonnaud, J.M. Jeschke, J.C. Russell, Invasion biology: specific problems and possible solutions, Trends Ecol. Evol. 32(1) (2017) pp. 13–22.*

On the other end of the spectrum are those who are critical of the concept of objective, value-free science. Science communication here has been influenced by the sociology of scientific knowledge, which has explored the social process by which scientific knowledge is created (e.g., Refs. [3,4]). In opposition to a positivist stance, there are those who adopt a constructivist position, approaching scientific knowledge from a postmodern perspective. At the extreme end of this perspective is a version of constructivism that denies the possibility of an objective reality. All understanding is constructed, bearing no relation to an external reality [5]. A more moderate version of constructivism holds that scientific knowledge is a product of social forces, yet still represents something "real" (Ian [6]). From a more moderate standpoint, scientific knowledge is accepted as contingent, but the value and applicability of scientific knowledge are still maintained.

For our purposes, it is worth noting the effect that different epistemologies have on approaches taken to science communication. Broadly speaking, those who adopt a constructivist position are more sympathetic to pluralism. In other words, constructivists will, to a lesser or greater extent, allow for a broader definition of what counts as "true" and what counts as "expertise." Those who adopt a positivist philosophy are more likely to embrace a top-down, deficit approach. This is because the idea that there are multiple valid perspectives is more challenging to maintain if one accepts the privileged epistemic status of scientific knowledge.

A criticism of constructivism could be that reporting science as "just" a social construct denies science its rightful place to define what is true. The view of science as facts, not opinion, can be seen in a wide range of science communication. As a recent example, in June 2017, the astrophysicist and popular science writer Neil deGrasse Tyson was interviewed on Radio 4. He said:

> … the only hope of a working democracy is to base laws and legislation on objective truth and science and its methods and tools is the best way we have ever come up with to establish what is objectively true in this world and what is objectively true is true whether you believe in it or not and the moment you start debating objective truth everybody's wasting everybody's time (deGrasse Tyson, BBC Radio 4 'Today' 8.6.17).

The beliefs that science is the best way of knowing about the world and that there are clear distinctions between fact and opinion, science and nonscience, have meant that science communication often emphasizes education. An assumption of this approach is that negative attitudes are due to ignorance. Sociological research often contends with the belief, implicit or explicit, that negative attitudes toward science can be ameliorated with more knowledge. Duncombe [7] argued that many academics (including scientists) share an enlightenment faith, a belief that if reasoning people simply have access to the truth, "the scales will fall from their eyes and they will see reality as it truly is and, of course, agree with us" [7,p. 7].

As noted above, the preferred model for science communication is that of "dialogue." A question thus emerges: How do you have a "dialogue" when you believe that one view is right (scientific) and another wrong (nonscientific)? One response to this question is to dissolve or blur the boundaries between expert and nonexpert. Michael [8], for example, states, "there is no easy differentiation between the expert and the popular, between the scientific and the lay, between the factual and fictional" [8,p. 370].

Prior [9] noted that this is often the tactic adopted in the sociology of health and illness. There has been a general trend to reduce this gap between "lay" knowledge and expertise.

Prior noted how this can be tracked in the language used. Language has slowly developed, moving from lay "beliefs" to lay "knowledge" to lay "expertise."

The assumption made in this chapter (i.e., our philosophical stance) is that there is a distinction to be made between expert and lay knowledge. This distinction may not be simple or binary, with boundaries blurred. Nevertheless, this distinction can still be useful. This position accepts that we can adopt clear epistemic values as a scientific or medical community. We can use these to evaluate the merits of particular views. This means that people can hold, to a lesser or greater extent, "correct" or "incorrect" views, at least about science. From this perspective, one cannot solve deficit theorizing by dissolving the differences between expert and nonexpert positions.

As such, understanding how different publics' views do not align with scientists can be appropriate. For example, it is important to know whether people believe that global warming is not real or that MMR vaccines cause autism. There is value in maintaining that scientific and medical communities have epistemic values and use these to evaluate the merits of particular views. We can see this in the context of genetics, where it is possible to see the value of distinguishing between different types of knowledge. For example, there is a vast array of everyday understandings about inheritance [10]. These may differ from scientific explanations, such as Mendelian inheritance patterns. If these beliefs conflict, this may present a challenge. For example, a person might have a belief, generated from their family experience, that they are not at risk for a genetic condition as they "take after" or even look like particular unaffected family members. In this situation, there is an extent to which a "deficit" approach is appropriate. In the example outlined above, it would be reasonable, and likely helpful, to assess a patient's misconceptions, distinguishing between medical knowledge based on scientific research and "everyday" knowledge.

3 Scientific truth exists, so why no deficits?

If we have this view of science, then it is reasonable to ask why a deficit approach is not appropriate. Why not conduct research to find out people's misconceptions and then design and enact education programs that address these? This may feel like common sense. However, there are good reasons for not taking this approach. We name four below:

- Deficit models do not account for how science communication is situated in broader contexts. It is not neutral. You cannot just "communicate the facts."
- Deficit models make it harder to be inclusive and see our own biases.
- Deficit models do not consider the complexity of the relationship between knowledge and attitudes.
- Deficit models do not consider the real-life, everyday ways in which people make sense of the world.

3.1 Historical context

Point one above can be demonstrated clearly by taking a historical view. This historical context in fact helps on two fronts. First, it allows a better understanding of how and when the shift from deficits to dialogues came about. Second, this historical perspective clearly helps to show how science communication has always been embedded within other strategic goals (social, political, and cultural). Science communication that aims to educate people about "just" the science (just the facts!) is based on an unachievable idea. To return to the key point of this chapter, science communication is never neutral.

3.2 Charles Babbage

Almost 200 years ago in 1830, the mathematician Charles Babbage published his work *Reflection on the Decline of Science in England.* Babbage wished for a greater professionalization of science. He was also concerned about the role of science in government. In short, Babbage wanted greater resources for science and a greater role for science in shaping policy. In the above work, Babbage blamed a lack of public interest for the decline of science. It was a view shared by some colleagues within the Royal Society.

A colleague wrote to Babbage in 1830, "It is a disgrace to men of science, and to the Royal Society, that they have not combined in a vigorous attempt to raise public feeling on the subject." Morrell and Thackray [11,12,p. 41].

In response to this, in 1831, the British Association for the Advancement of Science (BAAS) was formed "for reviving science in England." Two years later, William Whewell coined the term "scientist" to refer to this emerging professional group.

Thus, in the 1980s, science communication activities were embedded in this wider context (i.e., the professionalization of science). Public science was intended to bolster the authoritative role of science and scientists in shaping society. However, there was also a broader context. The BAAS organized great spectacles and demonstrations of science. These were intended to show the power and majesty of science. This, in turn, would demonstrate the technological and moral superiority of Britain and its empire. So, from the outset, even from the first moments when the term "scientist" came into use, we can see that science communication was inseparable from cultural, professional, and political contexts.

Further historical evidence shows how science communication was embedded in wider cultural settings. For example, 1959 saw the publication of Darwin's *Origin of Species.* It is impossible to divorce science communication activities that followed this from the social and religious context. Darwin's work was perceived as a challenge to religious authority. The relationship between evolutionary theory and religion is very complicated. It is not always the simplistic narrative that is often presented [13]. However, for our purposes, we can note how boundaries of scientific authority were negotiated in public through science communication. Public addresses and public lectures were a significant way in which scientists sought to promote Darwinism and impose their authority over religion. Scientists used public addresses and communication activities to fight for both cultural authority and financial resources [14].

The first half of the twentieth century saw radical shifts in how science was viewed by the public. Increased mechanization had displaced large numbers of people from their jobs, and both world wars highlighted the destructive power of science. On the other hand, there were causes for celebration. For example, this period saw huge advancements in medicine, such as the discovery and mass production of penicillin. A comprehensive history is not possible here. However, we can note that (as ever) science communication was inherently tied up with the broader social, political, and cultural contexts. For example, the science of inheritance could not be divorced from the Eugenics movement, which saw its appalling culmination with the Holocaust. We can see this in a children's picture book published by UNESCO in 1952. The book is called *What Is Race? Evidence from Scientists*. This book presents genetics as the route to an enlightened, scientific, non-prejudiced understanding of race [15]. Science communication here can be seen as having the strategic aim of distancing genetics from eugenics and preserving scientific authority and neutrality in the postwar period.

It is worth noting that the first half of the twentieth century saw an increasing prevalence of science fiction. Perhaps in response to the ambivalent impact of scientific advances, science fiction emerged as an important forum for public discussion and debate. For example, the influential biologist J.B.S. Haldane set out his view of how humans could control their own evolution in his short book *Daedalus; or, Science and the Future* [16]. Aldous Huxley wrote *Brave New World* [17] as a direct response to and satirization of Haldane's work.

3.3 The cold war

1957 saw the launch of the Soviet satellite Sputnik. When the Soviets took the lead in the space race, this led to anxiety in America about the decline of their scientific prominence. In a similar way to Babbage about 150 years before, many saw the solution to the decline of science as better science communication [18]. The overall aim was for better public knowledge and appreciation of science. It is in this Cold War period that we see the creation and utilization of the term *science literacy*. Increasing or improving science literacy is often stated as a reasonably neutral (at least apolitical) and self-explanatory goal. This is particularly relevant for genomics professionals, as science literacy has been the basis of *genetics literacy*. This has been used as a measure in a wide range of research settings, stated as something that needs to be improved in the public at large. However, as we can see, science literacy was not just about better knowledge. It was about improving science knowledge directed toward a political goal (i.e., beating the communists). Understanding these political goals is important, as they will shape how members of the public perceive, understand, and react to science communication efforts.

3.4 1985–2000 deficits to dialogues

In 1985, the Royal Society produced an influential report known as the Bodmer Report. As in the late 1950s in America, science in the United Kingdom in the mid-1980s was perceived to be in a state of decline. The Bodmer Report made a familiar diagnosis—lack of public knowledge and appreciation of science. Following the publication of the Bodmer Report, the key goal in science communication became the *public understanding of science*. This was seen as a key strategy to rejuvenate science. However, 15 years later, we saw quite a remarkable change of tone, at least in the United Kingdom. The "deficit model," which we have seen for so much of the history outlined above, had come under sustained criticism. A key report by the House of Lords Select Committee on Science and Technology [19] responded to this. From this point onward, science communication saw a radical shift. Public understanding of science (i.e., deficit models) became passé. Instead, terms such as public engagement, dialogue, and "science and society" were now preferred. The definitions of "dialogue" are varied [20]. However, the key idea of "dialogue" models is that science communication is not one way. The public cannot be seen as an empty vessel, to be filled with scientific knowledge.

This change in strategy was initiated in part due to issues of trust, especially in response to how science and government were seen to operate together. This trust was perceived to have been damaged due to events such as the controversy over genetically modified (GM) foods and the UK government's response to the BSE crisis. We can see this shift in tone in an expert from the report, which states, "Today's public expects not merely to know what is going on, but to be consulted; science is beginning to see the wisdom of this, and to move 'out of the laboratory and into the community' to engage in dialogue aimed at mutual understanding. Several of our witnesses agree that a shift along these lines is taking place" (House of Lords Select Committee on Science and Technology, 2000, p. 37).

We have only been able to briefly sketch the history of science communication. However, even this short historical account illustrated two important points. First, it shows the historical background of the shift from deficits to dialogues. Second, it allows us to see how science communication is trying to achieve something. Political, professional, cultural, and social aims are embedded within all forms of science communication.

3.5 Deficit models make it harder to be inclusive and see our own biases

Above, we outlined four reasons why deficit models are undesirable. The second of these is that when operating within a deficit model, it becomes harder to be inclusive and to see our own biases. Scholarship from the philosophy of science demonstrates that science itself can be deeply structured by the values and interests of its makers; scientific practice, even at its most rigorous, is not always or automatically self-correcting [21]. Science communication too often reflects the shape, values, and practices of dominant groups at the expense of the marginalized [22]. Science communication practices have been described as "sharply unevenly distributed" [23,p. 134]. White, middle-class people living with their families are more likely to take part in a range of activities including science talks, science festivals, museums, aquaria, and botanic gardens [24,25]. People from socioeconomically disadvantaged backgrounds are less likely to participate in science communication activities [26,27].

Importantly, questions about whose values, knowledge, and culture are reproduced in scientific knowledge and science communication are concealed (or at least minimized) if science is simply viewed as the objective truth. In deficit models—where science is regarded as ostensibly acultural—these dynamics are far more likely to be hidden or denied. This is a corollary of a positivist epistemology: Science cannot reflect values if it is viewed as *just* the objective truth. As such, science is valued not as a form of culture, but for its intrinsic value, that is, the power it has to explain the material world. To take the above quote from deGrasse Tyson again, "Science and its methods and tools is the best way we have ever come up with to establish what is objectively true in this world and what is objectively true is true whether you believe in it or not."

The second point then is that deficit models make it harder to communicate science in a way that is inclusive and allows us to see our own biases.

3.6 Deficit models do not consider the complexity of the relationship between knowledge and attitudes

The third point above is that deficit models often assume a positive correlation between knowledge and attitudes. However, more knowledge does not always lead to more positive attitudes toward science. Indeed, some research has shown that with controversial topics, more knowledge leads to more polarized views [28]. When there is a correlation between more knowledge and a more favorable view of science, this correlation is often weak [29].

Following this, it is important to recognize that resistance to science is not due to ignorance, but from the fact that science is a form of elite culture. We will discuss this in detail in the next section. However, the key aspect to note here is that people are not put off culture because they do not know enough. Often, it is a sense that something is just *not for them*. Because of this, deficit approaches (that assume ignorance) can often backfire.

3.7 Deficit models do not consider the ways people actually make sense of the world

The final point regarding deficit models is that they do not capture the myriad legitimate ways publics can engage with and question science outside of its own terms. Numerous authors have argued that the views of different publics regarding science are valid and legitimate. These views are based on personal experience, culture, or conventional wisdom. These "lay theories" enable people to reason and talk about complex science and to debate science in their own familiar terms [30,31].

When making sense of science, the scientific facts are only one piece of the picture. The types of knowledge and reasoning that people (including "experts") use to make sense of genetics are highly eclectic and syncretic. Exploring how people make sense of genetic information therefore requires an understanding of the complex ways that people make sense of science in the context of their own knowledge and experience.

4 Bourdieu, cultural capital, and science as culture

An important point raised above is how science works as a form of culture. How culture is valued—and consequently, how culture works to include and exclude—was theorized comprehensively by the sociologist Bourdieu. Bourdieu was interested in the ways in which privilege is created and maintained. He theorized that this is done through various forms of capital. He identified four: social, economic, symbolic, and cultural. Our focus here is cultural capital.

Bourdieu theorized that cultural knowledge confers social status and power. There are two important aspects to this. First, this value is arbitrary. Second, this arbitrary nature is hidden. In other words, the value of culture is seen to be natural or innate, which gives it power. An example is useful here. Two children might enjoy stories about magical, heroic figures fighting each other. One is reading Greek mythology, while the other is watching Marvel's latest superhero film. Bourdieu was interested in why one pastime would be considered worthwhile and the other a waste of time. Bourdieu dismissed the idea that one is better than the other. Instead, the value we place on types of culture is arbitrary.

Greek mythology is not inherently better than superhero films. However, one *is* valued more—what you might say is seen as more "legitimate." This means you can *seem* more intelligent, worthier, etc., if you can quote from Greek mythology than if you can quote from *Iron Man*. This relates to Bourdieu's second point. For culture to have power, its arbitrary value must be hidden. In Bourdieu's analysis, this is a way in which privilege can be perpetuated. Along with economic capital (money) and social capital (contacts), parents can provide their children with access to valuable cultural capital. Access to certain cultures is privileged. These cultures are then viewed as innately better and thus have greater value. In particular, this value can be thought of as "exchange value." This means culture has value linked to financial rewards (e.g., social status, networking, and employability).

4.1 Science capital

The concept of science capital builds on these insights and is increasingly used to frame science communication activities. For example, the Science Museum in the United Kingdom uses the concept to structure its engagement activities. It has also been adopted by the UK government in their recent report on attitudes toward science (2020). Building on the work of Bourdieu, the concept of science capital draws our attention to science as culture. This is particularly helpful in highlighting why deficits are impractical. Again, an example is useful if we think about another form of elite culture, such as opera. Opera, like classical music as a whole, has a diversity problem. So, let us say someone wants to improve participation or attitudes toward opera. They explain their plan is to visit people who do not like opera and find how little they know. They would then tell them facts about opera with the aim of increasing participation and improving attitudes. We can instantly see how inadequate this strategy would be. Yet this is the exact mistake the deficit models make regarding science communication. One of the main reasons people do not like science is not because they do not know enough. Instead, it is because they see it as an elite form of culture that is not created for them.

4.2 Problems with science capital

While science capital is a useful theoretical framework, it does have two problems. The first is that Bourdieu was a cynic. In Bourdieu's framework, there is very little room for thinking about the ways in which culture can be genuinely meaningful. Bourdieu's logic (and by extension, the logic of science capital) is that of commodities and strategy. The science capital lens allows you to see the "exchange value" of culture. However, this type of "capital logic" does not let you understand how engaging with science can be a genuinely meaningful cultural experience.

A second problem with science capital is that we can very quickly slip back into deficits. This is often reflected in the language of "building" science capital. Take this quote from the Governments 2020 report into public attitudes to science.

> The science capital measure developed by Louise Archer and her team, currently based at UCL Institute of Education comprises several dimensions including: scientific literacy and qualifications; participation in informal science learning; family science connections; talking about science in everyday life; and feeling 'connected' to science. Government policy is to encourage building and enhancing science capital among the general public as a means of upskilling people, to support economic growth, ensure a supportive social context for science and technology, and to widen engagement in science across all social groups [32].

Above, the government's stated aim is that of "building, enhancing and increasing science capital." It is important to critically assess how far we have really moved on from deficits here. Instead, it is perhaps more accurate to say that one deficit has been replaced by another. A telling phrase in the section above is "talking about science in everyday life." The assumption is that we can "build" science capital by supporting people to talk science in their everyday life. We can think about this phrase in relation to genetics and genomics. Presumably, building science capital would allow people to talk about genomics more in their everyday lives. Certainly, we know that not many people do talk about genetics and genomics in everyday life. However, many people find the concept of inheritance fascinating. While they might not talk about genomics per se, they talk about this in "everyday" ways such as discussing family stories and pop culture [33]. As such, the idea of "building science capital" could easily be read as "talking about science in the way we (as experts) want you to, in the correct way." The imposition of meaning is top-down and unidirectional. This is a key signifier of a deficit model.

5 Funds of knowledge

One way forward is to draw on the everyday ways in which people engage with science. A helpful framework for this is called "funds of knowledge." The funds of knowledge framework was developed with teachers and academics working in schools on the US–Mexican border. Historically, in the late 1980s, it was known that there was a persistent education gap between Mexican children and their peers. Interventions that had focused on remedying the perceived education deficits had not been successful. As such, educators worked with anthropologists, who went into the homes of students and worked with the families of schoolchildren. The aim was to find out what they were good at, what they knew about, and what they felt confident with—their funds of knowledge. They would then bring these into the classroom to improve the educational experience. As Gonzalez, Moll, and Amanti [34] put it, "The concept of 'funds of knowledge' is based on a simple premise: people are competent and have knowledge, and their life experiences have given them that knowledge."

A funds of knowledge approach allows one to genuinely move outside of deficits. As mentioned previously, two identified funds of knowledge, in relation to genetics, are family narratives of inheritance and pop culture [33]. However, we end up back at a similar point with regard to deficits and scientific truth. That is, what should be done when an individual's funds of knowledge contradict scientific views? For example, people often have lay theories of inheritance based on their familial experience. But what should be done if that directly conflicts with, say, Mendelian inheritance?

6 Hybrid space

This is where the concept of hybrid space can be helpful. Recent scholarship has used the funds of knowledge framework in conjunction with hybridity theory, developed by Bhabha [35]. Here, research acknowledges both expertise and the varied complex resources that people are able to draw on in a globalized environment. The third space is conceived of as an environment that brings together discourses from different communities and blends them to create new ideas and understandings [36].

Everyday (first space) and academic (second space) knowledges are often presented as oppositional. Moje et al. [37], however, describe the third space as being "in between" (1) different types of knowledge, creating new understanding as people blend everyday and academic knowledge. They say, "What seems to be oppositional categories can actually work together to generate new knowledges, new discourses, and new forms of literacy" Moje et al. [37].

There are a number of key aspects of hybrid space. The first is that it is about allowing different viewpoints or voices to be heard. Creating these opportunities is about power. Sometimes, it is literally about how physical space is used. Classroom setup or clinic setup, for example, can clearly signify whose views count and whose do not.

Hybrid space is also about allowing tensions to exist. Each "side," as it were, is committed to learning something new. So, to give an example from genetic counseling, somebody might say that they are not worried about the results of a genetic test, because they take after their father, whereas the faulty gene comes from their mother's side. This provides not a barrier but an opportunity to find out the meaning of the genetic test, how the different results will be interpreted, and what their significance would be for the patient.

7 Lessons for genomics education

This chapter ends with a summary of what we can learn from science communication for genomics education. For the first point, we return to the beginning of the chapter where we noted that science communication always tries to achieve something broader than the simple communication of the facts. This first point may not sit well for some, especially those who may feel their view of science fits closely with a "positivist" standpoint. It can feel manipulative, as if there is some sort of ulterior motive. Indeed, there is a comforting neutrality when one believes they are just communicating the facts (or believes that is what you are doing). We can perhaps see that this is the idea of being "non-directive" in genetic counseling. However, neutrality is not possible.

This can be seen in a positive light in that it provides an opportunity for dialogue. This insight is particularly helpful in situations with people or groups who are resistant or cynical about science. Neutrality is not possible, so it is to be abandoned. Instead, it is important to understand the wider context. If we learn about a person's or groups' political or religious beliefs, their cultural practices, their social situation, etc., we can begin to understand what tools they have to interpret the scientific information we are giving them. Instead of asking how we persuade people that science is correct, this approach leads us to ask questions such as: How might my message be interpreted? What might this person think I am trying to achieve? What do they think my motives are? Is this person likely to trust me? We can think of these questions as part of creating a "hybrid space," allowing different viewpoints to come together. These kinds of questions—which assume science communication is allowed to be multifaceted—are a good starting point for developing a science communications strategy that is based on dialogue.

7.1 Science is culture

Science itself is culture. Examples include science museums, science documentaries, celebrity scientists, science as a school subject, and science in film and TV. Culture is powerful when it is privileged. One of the ways it is privileged is through an esoteric set of rules. Opera, modern art, rap music, comic books—they all have their own tacit rules. When you know the rules, you can feel that this culture is "for you" and that you belong. If you do not know these rules, you can feel ostracized; this is not "for the likes of me." Science is no different in this respect. We saw above how questions about whose values, knowledge, and culture are reproduced in scientific knowledge and science communication are concealed (or at least minimized) if science is simply viewed as the objective truth. Viewing science as culture allows it to be critical and self-reflective without undermining the value of scientific knowledge.

Understanding this can help, particularly when thinking about our language. We may be driven by a desire to be as precise as we can with our language and to use the "correct" scientific terms. However, these terms have cultural weight as well as scientific meaning. When we use scientific terminology, we are being accurate, but we are also signaling that we are healthcare professionals and/or scientists and *we know these rules*. In some respects, it is like knowing not to applaud between movements when attending a classical music concert. Second, in clinical settings, we often want to create an environment where patients can clarify, question, or even challenge what we say. This is very difficult as they feel the "power" of science as culture—that science can only be questioned on its own terms.

This does not mean, of course, that we can never use scientific language. However, we need to be aware of what it means when we do. So, we can use this to develop guidelines such as only using complex scientific terminology when everyday words will not do (most of the time, everyday words work better) or to always have in mind that scientific language carries cultural significance. Bourdieu [38] showed that much of cultural power and value is arbitrary. When we are open about this, culture loses much of its power, at least its power to exclude.

8 Summary

How we communicate: funds of knowledge and hybrid space

We noted above that one of the failures of deficit models is that they often fail to account for the ways people make sense of the world. People make decisions by drawing on a range of different knowledges and experiences. It is syncretic. This is true for both laypeople and experts. How does one account for this and hold to the importance and value of scientific truth? The concept of hybrid space provides one answer to this. In a hybrid space, different points of view and different ways of understanding are not just welcome, but crucial.

In some sense, these different points of view can work to aid scientific understanding. For example, "everyday" forms of knowledge (or funds of knowledge) can work as "navigational" tools. In this sense, creating hybrid space can help people navigate complex or unfamiliar discourses such as medical texts and scientific literature. For example, people can give colloquial names to familial pathogenic variants (e.g., "Dad's wonky gene").

Allowing these into the discourse provides a rich opportunity for helping people navigate complex and new knowledges.

Hybrid space can also work as a "scaffolding" space, where people learn about how science can be linked to other parts of their lives. We can think back to the above government report of attitudes toward science. This report states the goal of building science capital. Part of this is enabling people to have "everyday conversations about science." From a hybrid space point of view, there would be a different focus. Instead, the aim would be for scientists and science communicators to better learn about people's lives. Through this, they would discover how science is relevant to the everyday conversations that people are already having.

Hybrid space can also be useful when thinking about what to do when we encounter viewpoints that are directly oppositional to scientific knowledge. Often when people engage in science communication and encounter an anti-scientific view, what they want, in essence, is surrender. They want someone with the alternative (anti-science) point of view to hold up their hands and say, "Yes, of course, you were right all along! Of course, vaccines work! Of course, climate change is real! Of course, wearing masks reduces the transmission of COVID-19! Of course, taking after my mother doesn't make it more likely I will inherit the BRCA1 gene fault from her!"

However, in a hybrid space, the idea is that multiple and different viewpoints can coexist. Tensions provide opportunities for learning. We allow different funds of knowledge into the conversation. Tensions are allowed to exist. They are explored. Funds of knowledge are talked about with an engaged curiosity. Through this, new knowledge is created. Both sides (hopefully) learn. This commitment to hybridity will, in the long term, hopefully facilitate a more fruitful dialogue, *genuine dialogue.*

This chapter has provided some different ways of thinking about science communication. The aim has been to use scholarship from science communication to provide a framework for how genomics professionals can approach their own communication with patients, colleagues, and lay audiences.

Acknowledgments/disclaimer

The authors would like to acknowledge the support of Wellcome funding [206194] paid to Society and Ethics Research Group, Wellcome Connecting Science, Cambridge, and also the doctoral support from Louise Archer and Jennifer DeWitt. The authors would also like to declare no conflicts of interest in relation to this work.

References

[1] S.J. Lock, Deficits and dialogues: Science communication and the public understanding of science in the UK, in: Successful science communication: Telling it like it is, 2011, pp. 17–30.

[2] A.F. Chalmers, What Is this Thing Called Science?, Hackett Publishing, 2013.

[3] K.D. Knorr-Cetina, The micro-sociological challenge of macro-sociology: Towards a reconstruction of social theory and methodology, in: K.D. Knorr-Cetina, A.V. Cicourel (Eds.), Advances in Social Theory and Methodology: Toward an Integration of Micro-and Macrosociologies, Routledge, 1981, pp. 1–47.

[4] B. Latour, S. Woolgar, Laboratory Life: The Social Construction of Scientific Facts, 1979. Beverly Hills.

[5] E. Von Glasersfeld, Radical Constructivism, Routledge, 2013.

[6] I. Hacking, The Social Construction of What?, Harvard University Press, 1999.

[7] S. Duncombe, Dream: Re-Imagining Progressive Politics in an Age of Fantasy, The New Press, New York, 2007.

[8] M. Michael, Comprehension, apprehension, prehension: heterogeneity and the public understanding of science, Sci. Technol. Hum. Values 27 (3) (2002) 357–378.

[9] L. Prior, Belief, knowledge and expertise: the emergence of the lay expert in medical sociology, Sociol. Health Illn. 25 (3) (2003) 41–57.

[10] K. Featherstone, P. Atkinson, A. Bharadwaj, A. Clarke, Risky Relations: Family, Kinship and the New Genetics, 2006.

[11] J. Morrell, A. Thackray, Brewster to Babbage, 16 Jun 1830 British library London fols 229-30, 1981, p. 50.

[12] J. Morrell, A. Thackray, Gentleman of Science: Early Years of the British Association for the Advancement of Science, Oxford University Press, Oxford, 1981, p. 41.

[13] M. Ruse, Can a Darwinian Be a Christian?: The Relationship between Science and Religion, Cambridge University Press, 2004.

[14] T.F. Gieryn, Boundary-work and the demarcation of science from nonscience: strains and interests in professional ideologies of scientists, Am. Sociol. Rev. (1983) 781.

[15] J. Bangham, What is race? UNESCO, mass communication and human genetics in the early 1950s, Hist. Hum. Sci. 28 (5) (2015) 80–107.

[16] J.B.S. Haldane, Daedalus, or Science and the Future: A Paper Read to the Heretics, Cambridge on February 4th 1923, E. P. Dutton & Company, New York, 1924.

[17] A. Huxley, 'Brave New World, Vintage Classics, 1932. ISBN 13 9780099518471.

[18] K. Lock, J. Pomerlau, J. Causer, D.R. Altmann, M. McKee, The global burden of disease attributable to low consumption of fruit and vegetables: implications for the global strategy on diet, Bull. World Health Org. 83 (2005) 100–108.

[19] House of Lords Select Committee on Science and Technology, 2000. https://committees.parliament.uk/committee/193/science-and-technology-committee-lords/publications/.

[20] S.J. Lock, Lost in Translation: Discourses, Boundaries and Legitimacy in the Public Understanding of Science in the UK, University of London, University College London, United Kingdom, 2008.

[21] A. Wylie, L.H. Nelson, Coming to Terms with the Value (s) of Science: Insights from Feminist Science Scholarship, 2007.

[22] E. Dawson, Reimagining publics and (non) participation: exploring exclusion from science communication through the experiences of low-income, minority ethnic groups, Public Underst. Sci. 27 (7) (2018) 772–786.

[23] K. Rommetveit, B. Wynne, Technoscience, imagined publics and public imaginations, Public Underst. Sci. 26 (2) (2017) 133–147.

[24] M.O.R.I. Ipsos, Public Attitudes to Science 2011, Department for Business Innovation and Skills, London, 2011.

[25] M.O.R.I. Ipsos, Public Attitudes to Science 2014, Department for Business, Innovation and Skills, London, 2014.

[26] E. Dawson, Reframing social exclusion from science communication: moving away from 'barriers' towards a more complex perspective, J. Sci. Commun. 13 (2) (2014) 1–5.

[27] E. Dawson, Social justice and out-of-school science learning: Exploring equity in science television, science clubs and maker spaces, Sci. Educ. 101 (4) (2017) 539.

[28] D.M. Kahan, E. Peters, M. Wittlin, P. Slovic, L.L. Ouellette, D. Braman, G. Mandel, The polarizing impact of science literacy and numeracy on perceived climate change risks, Nat. Clim. Chang. 2 (10) (2012) 732–735.

[29] N. Allum, P. Sturgis, D. Tabourazi, I. Brunton-Smith, Science knowledge and attitudes across cultures: a meta-analysis, Public Underst. Sci. 17 (1) (2008) 35–54.

[30] B. Wynne, Misunderstood misunderstanding: social identities and public uptake of science, Public Underst. Sci. 1 (3) (1992) 281–304, https://doi.org/10.1088/0963-6625/1/3/004.

[31] E.K. Nisbet, J.M. Zelenski, S.A. Murphy, The nature relatedness scale: linking individuals' connection with nature to environmental concern and behavior, Environ. Behav. 41 (5) (2009) 715–740.

[32] BEIS Research Paper Number 2020/012, Accessed November 10th 2020 at https://assets.publishing.service.gov.uk/government/uploads/system/uploads/attachment_data/file/905466/public-attitudes-to-science-2019.pdf.

[33] J. Roberts, L. Archer, J. DeWitt, A. Middleton, Popular culture and genetics; friend, foe or something more complex? Eur. J. Med. Genet. 62 (5) (2019) 368–375.

[34] N. González, L.C. Moll, C. Amanti (Eds.), Funds of Knowledge: Theorizing Practices in Households, Communities, and Classrooms, Routledge, 2005.

[35] Homi K. Bhabha (1994) The Location of Culture. Routledge; ISBN 9780415336390.

[36] E.W. Soja, Thirdspace: Journeys to Los Angeles and Other Real- and-Imagined Places, Blackwell, Malden, MA, 1996.
[37] E.B. Moje, K.M. Ciechanowski, K. Kramer, L. Ellis, R. Carrillo, T. Collazo, Working toward third space in content area literacy: an examination of everyday funds of knowledge and discourse, Read. Res. Q. 39 (1) (2004) 38–70.
[38] P. Bourdieu, Distinction (R. Nice, Trans.), Harvard, University Press, Cambridge, MA, 1984.

Further reading

[1] F. Courchamp, A. Fournier, C. Bellard, C. Bertelsmeier, E. Bonnaud, J.M. Jeschke, J.C. Russell, Invasion biology: specific problems and possible solutions, Trends Ecol. Evol. 32 (1) (2017) 13–22.
[2] E. Dawson, When science is someone else's world, in: L. Avraamidou, W.-M. Roth (Eds.), Intersections of Formal and Informal Science, Routledge, London and New York, 2016, pp. 82–92.
[3] D. Dennett, Guardian Interview Found at 12 Feb 2017, 2017. https://www.theguardian.com/science/2017/feb/12/daniel-dennett-politics-bacteria-bach-back-dawkins-trump-interview.
[4] S.J. Lock, Cultures of incomprehension?: the legacy of the two cultures debate at the end of the twentieth century, Interdiscip. Sci. Rev. 41 (2–3) (2016) 148–166.
[5] M.C. Nisbet, A. Dudo, Entertainment media portrayals and their effects on the public understanding of science, in: Hollywood Chemistry: When Science Met Entertainment, American Chemical Society, 2013, pp. 241–249.
[6] M.C. Nisbet, R.K. Goidel, Understanding citizen perceptions of science controversy: bridging the ethnographic—survey research divide, Public Underst. Sci. 16 (4) (2007) 421–440.

Global perspectives of genomic education and training

Dhavendra Kumar

William Harvey Research Institute, Bart's and The London School of Medicine & Dentistry, Queen Mary University of London, London, United Kingdom

1 Introduction

Scientific and technical education combined with research and development are the main drivers for globalization of any branch of science and technology. Among the biological and life sciences, genetics and contemporary genomics hold a significant status. Several universities and academic institutions are actively engaged in delivering and enhancing the status of genetic or genomic education at all levels. Core and advanced elements of genetic and genomic sciences feature in most curricula and syllabuses of secondary education, graduate courses, postgraduate diploma and master's courses, and several adult and continued education courses. These can be conveniently searched and catalogued through Google Scholar (www.scholar.google.com) and several other online sources.

While the momentum and quality of global genetic and genomic education have continually expanded in the developed nations, several countries in the underdeveloped and developing nations are unable to invest and offer even basic levels of life sciences education, particularly related to genetics and genomics. Inevitably, huge gap exists in both quantitative and qualitative parameters. However, to some extent, this gap is gradually narrowed down with the help of international charitable organizations, notably the Wellcome Trust (www.wellcome.ac.uk), the Rockefeller Foundation (www.rockfellerfoundation.org), and the Gates Foundation (www.gatesfoundation.org). In addition, there are several United Nations educational programs harnessed by many less developed nations. The United Nations Educational, Scientific and Cultural Organization (www.UNESCO.org) supports and collaborates with many global projects that include genetic and genomic education, such as the Human Variome Project discussed later in this chapter (www.humanvariome-project.org). Similarly, several streams of scholarships and training fellowship exist funded by the Commonwealth (www.commonwealtheducationtrust.org), the International development projects of the United Kingdom government (www.gov.uk), and the British Council (www.britishcouncil.org). Leading international agencies like the Organization for economic and community development (www.OECD.org) have provided information and critical data for member nations to formulate the strategy and incorporate genetic and genomic sciences and related biotechnical development projects.

The main purpose of this chapter is to introduce global perspectives on genetic and genomic education and training. Emphasis is given to the most relevant organizations and programs that focus on specific skills and competencies essential for the infrastructure and delivery of healthcare services that directly or indirectly include core or specialized genetic and/or genomic elements.

2 Global organizations and genomic education

2.1 Human Variome project

The Human Variome Project is an international nongovernmental organization that is working to ensure that all information on genetic variation and its effect on human health can be collected, curated, interpreted, and shared freely and openly (www.humanvariomeproject.org). It was founded in 2006 by the Late Professor Dick Cotton in Melbourne, Australia. It is truly a global initiative sponsored by around 80 nations and affiliated with several international organizations, including World Health Organization, OECD, European Commission, United Nations Educational, Scientific and Cultural Organization (UNESCO), March of Dimes Foundation, Centers for Disease Control and Prevention, and others. The whole consortium is commonly referred to as Global Variome.

It has a central coordinating function for national and international efforts to integrate the collection, curation, interpretation, and sharing of information on variation in the human genome into routine clinical practice and research. The consortium includes around 1000 individual researchers, clinicians, healthcare professionals, policy makers, and organizations from 81 countries that collaborate to develop and maintain the necessary standards, systems, and infrastructure to support global-scale genomic knowledge sharing. The project is essentially as advisory group advising on the development and operation of physical data storage and sharing infrastructure. This remains the responsibility of international disease groups, individual national health systems, and individual members.

The Human Variome Project undertakes the following tasks:

- Collaboratively developing technical standards and harmonized, common approaches so that data from different sources can be easily shared in an interoperable manner that is sensitive to the ethical, legal, and social requirements of both the data sources and consumers;
- Coordinating an international platform to facilitate discussion of genomics in global health with the aim to foster necessary professional interaction and debate in the area of genomics, global health, service delivery, and safety;
- Linking world leading professionals and institutions with genomics professionals, researchers, and academics in all parts of the world, facilitating knowledge exchange and interactive debate;
- Establishing a global evidence base for knowledge sharing in medical genetics and genomics and bringing relevant issues to the attention of Ministries of Health, Science and Technology, and Education.

The governance of this organization is managed by the International Co-ordinating Office that is linked to individual country's base. There are three specific subcommittees that guide and provide regular inputs on matters related to country node affairs (International Confederation of Countries Advisory Council), scientific matters (International Scientific Advisory Committee), and gene/genome database development (Gene/Disease Specific Database Advisory Council). There are global projects that offer opportunities for acquiring and/or enhancing skills and competencies (Global BRCA Challenge, Global globin 2020, Global Familial Heart Challenge). Recently, the project has agreed to work closely with other global organizations (Human Genome Organization International—HUGO— and Human Genome Variation Society—HGVS) for increasing efficiency and maximizing benefits of rapid genomic knowledge and advances.

2.2 Human genome organization international (HUGO)

The Human Genome Organization (HUGO) is the international organization of scientists involved in human genetics (www.hugo-international.org). It was founded in 1988 at the first meeting on genome mapping and sequencing at Cold Spring Harbor Laboratory, Long Island, NY. Currently, it has around 2000 members from 87 countries. HUGO has, over the years, played an essential role behind the scenes of the human genome project. With its mission to promote international collaborative effort to study the human genome and the myriad issues raised by knowledge of the genome, HUGO has had noteworthy successes in some of the less glamorous, but nonetheless vital, aspects of the human genome project.

HUGO has continually worked focusing on the medical implications of genomic knowledge. Moving forward, HUGO is also working to enhance the genomic capabilities in the emerging countries of the world. The excitement and interest in genomic sciences in Asia, Middle East, South America, and Africa are palpable, and the hope is that these technologies will help in national development and health. Notable contributions of HUGO include genetic and genomic education resources, gene nomenclature, deciphering gene/genome variants, global human genomics knowledge exchange platform, and interaction with major human genomics projects.

2.3 Global alliance for genomics and health (GA4GH)

The **Global Alliance for Genomics and Health** (GA4GH; www.ga4gh.org) is an international, not-for-profit profit alliance of around 2000 members from just over 600 organizations. The core aim is to collaborate, innovate, and accelerate the potential of genomic research and applications in medicine to advance human health. The alliance membership is drawn from healthcare, research, patient advocacy, life science, and information technology. The GA4GH community is working together to create frameworks and standards to enable the responsible, voluntary, and secure sharing of genomic and health-related data. There are multiple work streams led by many different driver projects. The genomic knowledge and skills feature across many work streams, particularly setting the genomic knowledge standards, clinical and phenotypic data capture, and regulatory and ethics. The alliance includes the **GA4GH Connect**, a 5-year strategic plan that aims to drive uptake of standards and frameworks for genomic data sharing within the research and healthcare communities.

2.4 Global genomic medicine collaborative (G2MC)

The Global Genomic Medicine Collaborative (G2MC; www.g2mc.org) is an organization that is creating a community of global leaders dedicated to advancing genomic medicine implementation in clinical care. G2MC is identifying opportunities to foster global collaboration to demonstrate value and the effective use of genomics in medicine. The G2MC was started as an "action collaborative" in the United States National Academies of Medicine and also was supported by the National Human Genome Research Institute of the NIH. Now as an independent not-for-profit charitable organization, it seeks to improve global health by catalyzing the implementation of genomic tools and knowledge into health care delivery globally.

The G2MC includes a number of working groups. The mission of the Education Working Group is to provide medical education on the implementation of genomic medicine in clinical practice. The group has a number of priority areas:

- Develop courses, case studies, assessment tools, and other educational resources. Inspire, engage, and involve young investigators in genomic medicine.
- Collaborate with other educational groups/organizations to develop educational resources and expand outreach opportunities.
- Develop programs that support the Flagship Projects

This global genomic medicine organization is specifically interested in utilizing the diversity of G2MC to accomplish the following:

- Conduct a needs assessment for genetics/genomics training in the developing world.
- Utilize the results of the needs assessment to develop a curriculum and tools for implementation and evaluation.
- Pilot and test the curriculum at G2MC members' sites.

2.5 The global genomics nursing alliance (G2NA)

The Global Genomics Nursing Alliance (G2NA; www.g2na.org) aims to accelerate the integration of genomics into everyday nursing practice. It was established in 2017 to initiate that effort, to promote nursing in genomics healthcare, and to accelerate the integration of genomics across everyday nursing practice. G2NA believes that by working together, we can mobilize information, resources, and strategies to realize the benefits of genomics for the patients and families that we serve. **The G2NA vision** is to serve as the unified international voice for advancing and integrating genomics into nursing practice, supporting nurses to realize their full potential through integrating genomics across nursing practice to improve healthcare for all.

2.6 Global consortium for genomic education (GC4GE)

The *Global Consortium for Genomic Education (GC4GE)* is a major new initiative led by the Genomic Medicine Foundation UK (www.genomicmedicine.org) to bring together many committed scientists, clinicians, professionals, institutions, laboratories, and organizations for developing a concerted program for genomic education and training delivered globally with a particular focus on developing and less developed nations. The main of the consortium is to organize, coordinate, and steer a global program for the integrated genomic education and training. The consortium has several objectives:

- To identify and engage with major professionals, institutions, and organizations involved with genetic/genomic education and training.
- To develop the core framework for the consortium in collaboration with participating groups.
- To develop a network for the consortium working on the "hub and spoke" model.
- To develop the bespoke integrated genomic education and training program incorporating the major components of genome science and technology.
- To develop mechanisms for delivering the integrated genomic education program on a regional and country basis.

The consortium (GC4GE) is structured on the "hub and spoke" model (Fig. 1). The central hub coordinates and manages the functioning of the consortium. Each spoke of the GC4GE represents a particular region or country; for example, Asia Pacific region will be one unit since it includes many small countries compared to India or China. These regions and countries are selected from less developed and developing nations category. Experience and the knowledge base from the developed nations hub will be used for developing the curriculum and assembling the faculty. The faculty works jointly with the respective regional and country faculty for developing bespoke program and assessing the outcomes.

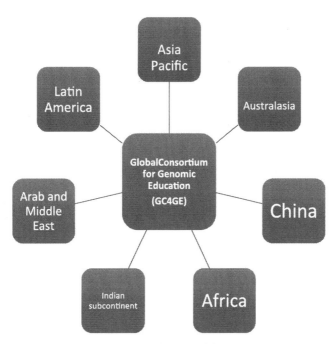

FIG. 1   Scope of the global consortium for genomic education (GC4GE).

In the first phase of development, priority areas of the GC4GE include:

1. Genome technology, including bioinformatics
2. Medical and health applications
3. Genomics and bio-economy
4. Ethics, legal, and social issues (ELSI)

High-caliber experts are convened to join the advisory and executive teams. The education and training program of each main sector is delivered through seminars, series of webinars, and training workshops. The individual educational activity is appropriately tailored for accreditation by the relevant regional and/or national authority. It is also envisaged that the educational program might also be approved for accumulated credits or affiliated with existing or new academic courses, such as the Post Graduate Diploma in Genomic Medicine and Healthcare (The University of South Wales) or MSc in Genomic Medicine (The Swansea University, South Wales, UK) and other similar courses offered by many institutions and universities in UK and other places. It is anticipated that the above programs will harness highly motivated and academically bright students for further higher studies in genome science and technology, including PhD-level research projects.

Success of the genomic education and training consortium depends on the support and participation of major international stakeholders and the network of participating regional and/or institutions and organizations. The Genomic Medicine Foundation works closely with other global agencies such as the March of Dimes Birth Defects Foundation, WHO South East Asia Region Office, Global Genomic Medicine Collaborative (G2MC), and Global Alliance for Genomics and Health (GA4GH).

3 Distance learning for genomic skills and competencies

Since the dawn of the millennium, rapid advances in digital technology and availability of progressively low-priced digital devices have brought the digital era to real existence. All over the world, children, adults, and elderly of all ages use every day one or more form of digital technology for communication, information, learning, entertainment, and social interactions. In addition, digital automation and artificial intelligence technologies have given us access to several avenues that were considered hugely expensive and unattainable.

Novel digital modalities now exist for learning and acquiring new skills. Most universities and academic organizations employ digital platforms and portals for teaching, continuous evolution of courses, and individual assessments. Distance learning is now an acceptable format. However, there are few governance issues, for example, authenticity and probity of students at the other end. To some extent, these are now managed with a rapid use of digital audio-video real-life online communication portals. The COVID-19 pandemic provided a unique opportunity that led to massive global use of these digital channels for teaching, practical skills learning, tutorials, assessments, and formal end of the course qualifying examinations. A complete overhaul of the whole education system is underway and likely to be with us for generations.

Sophistication of the information technology (IT) is of fundamental importance to the successful planning and delivering any distance learning program. This would require considerable investments in developing and commissioning at the institutional level with the full support of governmental and nongovernmental agencies. Special safeguards would need to be in place appropriately governed by statutory rules and regulations. Issues like probity, confidentiality, consent, and cyber safety would need to be carefully considered and specifically placed. Expectations of people and media would need to be taken into account and actioned. The Commonwealth Centre for Digital Health (www.cwcdh.com) is probably a good example that sets out different models and guidelines in keeping with broad societal and individual Nation's expectations from digital education systems, specifically the distance learning courses.

Constituent elements and drivers for genetic and genomic distance courses would require several specific sections. Each course would need to be supported by generous and uninterrupted access to online resources to the course curriculum guidance, books and journals, seminars and symposia, and genetic and genomic databases. The role of bioinformatics and computational biology is universally acknowledged for structuring and maintenance of key genotype-phenotype databases, for example, Online Mendelian Inheritance in Man (www.OMIM.org), Sanger Decipher database (www.deciphergenomiccs.org), Human Gene Mutation Database (www.hgmd.cf.ac.uk), ClinGen (www.clinicalgenome.org), and ClinVar (https://ncbi.nlm.nih.gov/clinvar/). Each institution or the responsible course director would need to ensure a safe and effective delivery of the course with robust and unambiguous fair evaluation and assessment systems. These would need to be tailored according to the nature of the course whether genetic or specialized subject. The respective regulatory or "watch dog" statutory authority would be responsible for setting up and executing necessary standardization and accreditation processes. In addition, ethical and legal issues would be governed by independent governmental or nongovernmental organizations; for example, in the United Kingdom, the Nuffield Bioethics (www.nuffieldbioethics.org) and Genetic/Genome Ethics (www.genethicsforum.ning.com) play a key role in genetic and genomic medicine-related matters, including teaching and training. Most of the major professional organizations, like the British Society of Genetic Medicine (www.bsgm.org.uk), European Society of Human Genetics (www.eshg.org), and the American Society of Human Genetics (www.ashg.org) offer guidance and assistance in setting up teaching and training short courses. The National Human Genome Research Institute (https://www.genome.gov) of the National Institutes of Health (https://www.nh.gov) has established dedicated distance learning modules for genomic medicine education for healthcare professionals and the public.

4 Global status of genetic and genomic education and training

Rapid globalization of medical teaching and training has led to several programs that design and deliver education and training in core and specialist sectors of different medical, dental, nursing, and healthcare disciplines. Medical genetics and genomics as a specialist field is yet to be recognized by most countries. However, structured and validated programs exist that offer strong knowledge base along with opportunities to acquire and enhance skills and competencies. The global distribution (see Fig. 2) covers all continents, however with relatively higher proportion in developed nations.

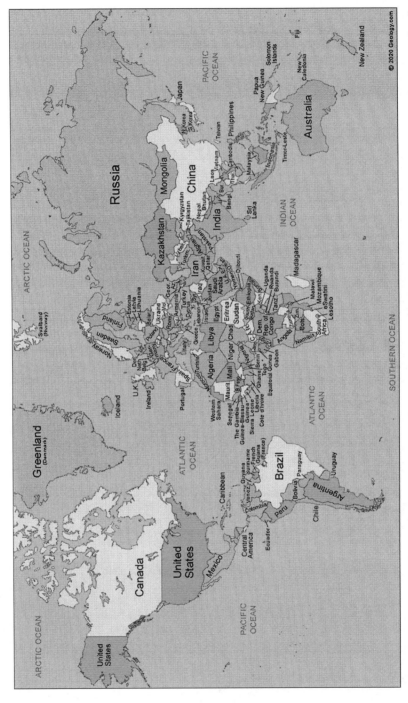

FIG. 2 Global map showing key medical genetics and genomics teaching and training programs and courses—Canada, USA, Mexico, Brazil, Argentina, United Kingdom, European Union (limited), Russian Federation (limited), Egypt, South Africa, Iran, UAE, India, Sri Lanka, China, Taiwan, Japan, Philippines, Malaysia, Singapore, Australia, and New Zealand. *Image downloaded from google.com/courtesy Geology.com.*

4.1 United Kingdom and the European Union

Even before the advent of chromosomes and the DNA double helix, genetics and heredity were taught by many British and European universities and academic institutions. These were encompassed in graduate (BA or BSc), postgraduate (MA, MSc, or MPhil), and higher doctoral (PhD, DPhil) degrees curriculum. The current scenario includes several opportunities for graduate, postgraduate, continued adult education, and lay people courses. These could be divided into two separate categories as regards to knowledge, skills, and competencies. The list is too long to include in this chapter. Only medically relevant courses and training programs are discussed.

4.1.1 *Knowledge based-*

Graduate and postgraduate courses

There are currently several graduate (BSc Hons) and postgraduate Master's degree courses offered by many UK and European academic institutions. The curriculum includes basic and applied genetics with some laboratory genetics exposure. Typically, the Master's level course includes around 6 months' research-related teaching with the requirement of successful submission of a dissertation. Most successful students opt for further genetic and genomic education by research (PhD or DPhil) or enroll with serious professional medical (MB, MBBS, MD) or veterinary training (BVSc, BVetMed, BVCM&S, or BVMS).

Intercalated BSc during MB/MBBS

The option or opportunity to study medical applications of genetics during the undergraduate medical degree course (MB; MBBS) is available in few selected British medical schools. Similar arrangements also exist in other places. The Bachelor degree course is embedded with the medical curriculum. However, the student would need to step out of the course and thus effectively be 1 year behind the peers. The intercalated BSc in Medical Genetics is popular among many medical students. All students undergo formal teaching, usually as part of other postgraduate genetic teaching courses, for example, MSc in Genomic Medicine or MSc in Genetic Counseling. The one-year supervised placement includes short period shadowing genetic clinics, genetic cases meetings, working with genetic counselors, and laboratory exposure. Every student undertakes a small actively supervised project leading to dissertation. Some of the successful medical students with intercalated BSc in Medical Genetics might eventually select a professional career in Medical Genetics and Genomics.

MSc courses in United Kingdom

Master's level medical genetic courses in the United Kingdom (UK) have been taught for considerable number of years. These were revisited and restructured following the successful 100,000 Genome Projects. Currently, there are around 11 MSc in Genomic Medicine degree courses offered by various UK universities. These popular Master's courses are open to UK, European Union, and non-EU students. Some of the UK students are predominantly for the medical, nursing, and healthcare professionals working in the UK National Health Service (NHS). There are complex organizational and regulatory mechanisms in place managed by the Health Education England (www.nee.nhs.uk). Similar arrangements exist in UK devolved nations (Wales, Scotland, and Northern Ireland). A successful professional with MSc in Genomic Medicine usually goes forward for structured professional vocational training in a medical specialty, midwifery, specialist nursing, healthcare therapy, or genetic counseling. It is anticipated that a typical NHS professional practicing genetic or genomic medicine would have either MSc or equivalent higher qualification.

Postgraduate diploma in genomic medicine and healthcare

Most of the UK, European, and North American postgraduate courses in Medical Genetics or Genomics are taught curricula that require mandatory physical attendance. It might be unaffordable for many and difficult for personal career and family reasons, particularly for students from less developed nations in Africa, South Asia, and Asia-Pacific. With the advent of affordable online portals, distance learning courses have offered decent alternatives to physical-taught degree courses. The author set up and directed a postgraduate diploma course with similar objectives at the University of Wales in the UK (www.southwales.ac.uk). The one-year course curriculum was split in six distance learning modules, namely, applied basic cell biology and molecular genetics; clinical genetics; laboratory genetics and genomics; genetic counseling; fetal medicine and prenatal diagnosis; and ethical, legal, and social issues (ELSI). This course has helped many established and new emerging medical and nursing healthcare professionals to develop and establish genomic medicine service units in India, Africa, and Asia-Pacific.

European diploma in medical genetics and genomics

The European Board of Medical Genetics (www.ebmg.eu) has commissioned a Diploma course in Medical Genetics and Genomics. The course is regulated by the Section of Medical Genetics of the European Union of Medical Specialists (www.UEMS.eu). This knowledge-based diploma course is part of the 5 years' professional clinical genetics and genetic counseling vocational training program. The EBMG recommends that medical geneticists (MDs) should undergo training and education and be able to demonstrate competence in the areas defined by UEMS section of the clinical genetics curriculum.

Professional/accredited training

Historically, it took several years before medical genetics as a whole or its specific components (clinical genetics, genetic counseling, and laboratory genetics) were acknowledged and accorded the much-needed specialty status. In early years, spanning almost three decades, the teaching and training were available on random and ad hoc basis. There were no structured curricula and skills/competencies were judged on personal basis. Toward the second half of the 1980s, the UK and European Union developed few medical genetics training programs. The UK Medical Royal Colleges took charge of higher medical and surgical specialist training, including clinical genetics. The clinical genetics was then included by the Royal College of Physicians (RCPs) as a medical specialty. The laboratory genetics training program followed administered by the Royal College of Pathologists (RCPath). Initially, there was no clear distinction between genetic nursing and genetic counseling. Earlier training included elements of both aspects of genetic healthcare. Gradually, these got separated and genetic counseling got the much-needed recognition regulated by the dedicated professional organization, the Association of Genetic Nurses and Counselors

(AGNC), the constituent body of the British Society of Human Genetics (BSHG), later the British Society of Genetic Medicine (BSGM). Similar arrangements followed across the European continent, irrespective of the European Union (EU).

Higher specialist medical training in clinical genetics   Currently in the UK, the higher specialist medical training in clinical genetics and laboratory genetics is the responsibility of the Joint Committee for Higher Medical Training (JCHMT). The original JCHMT was constituted by the Federation of Royal College of Physicians (UK), which had the delegated responsibility for the administration of specialist training in general internal medicine and the medical specialties and subspecialties. The role of the JCHMT was taken over by the Joint Royal Colleges of Physicians' Training Board (www.JRCPTB.org.uk), which develops the frameworks for all three Royal Colleges of Physicians of the UK (Edinburgh, Glasgow, and London). Each specialty is served by the Specialist Advisory Committee (SAC), largely from the specialty with one representative from JRCPTB.

The specialist clinical genetics training program is structured for 4 years. Most trainees have minimum 3 years' postgraduate training in Pediatrics, Medicine, Obstetrics, and Gynecology. Some trainees came from general practice and occasionally surgical background, particularly from orthopedic surgery. The recently updated and revised clinical genetics curriculum includes fair proportion of genomic applications, particularly the diagnostic genomics (www. clingensoc.org and www.bsgm.org.uk).

Genetic counseling training program   The status of genetic counseling as an independent healthcare discipline remained unclear for several years. In most cases, new appointees got training and developed skills and competencies through learning on the job. Both in UK and EU, this situation continued for a considerable period. Eventually, short courses and full Master's level courses in genetic counseling were introduced. The MSc in Genetic Counseling offered by Manchester and Cardiff became very popular both across Europe and far-away places like South Africa and Australia. These were followed by similar courses in few EU nations. The structured knowledge-based courses combined with formal clinical training facilitated professionally recognizable registration as a genetic counselor (GC). There is strong political and professional commitment to develop a formal statutory regulatory system for GC registration and continued professional development (www.agnc.org.uk).

4.2 United States of America and Canada

The status of formal and semi-formal training of medical geneticists, including genetic counselors, is comparatively advanced in the United States of America (USA) and Canada. The medical genetics and genomics are constituted within the American Board of Medical Genetics (www.abmgg.org) and included as a medical specialty by the American Board of Medical Specialties (www.abms.org). The ABMG medical genetics certification commenced in 1990. The responsibility for medical genetics training and certification in Canada was taken up by the Canadian College of Medical Genetics (www.ccmg-ccgm.org). Both USA and Canada have laid out specific and explicit training and eligibility requirements for certification in different categories. All qualified and certified medical geneticists in any category are listed as "Board certified" and accordingly practice within the statutes of various states. There are clear guidelines for continued medical education and 5 yearly "re-certification."

The status of genetic counseling became clear when the American Board of Genetic Counseling was established in 1993 following the recognition of the ABMGG. However, since 1981 the genetic counseling certification was the responsibility of ABMG. Since 1993, ABGC has defined the professional standards of competence for the genetic counseling profession. The number of genetic counselors recognized by ABGC has risen from 495 to more than 5000, and the number of graduate programs has more than doubled. Certified genetic counselors continue to integrate counseling services into an increasing number of medical specialties, such as oncology, cardiology, and psychiatry. The eligibility to write the genetic counselor certification examination depends on the recommended period of training in an accredited course, usually Masters, offered by several USA and Canadian Centers. In Canada, genetic counselors are certified by the Canadian Board of Genetic Counseling-Conseil Canadien de Conseil Génétique (CBGC-CCCG). Prior to 2021, the certification process was overseen by the Certification Board of the Canadian Association of Genetic Counselors. The genetic counseling profession is currently unregulated in Canada, and as such, practitioners are not governed by provincial and territorial legislations, which ensure safe, competent, and ethical practice in the interest of public protection. The national certification credential is an important basis for the evolution of professional legislation and regulation in Canada (www.cagc.accg.ca).

In addition to the certification programs by the respective North American medical specialty organizations, several universities offer a number of academic and research qualifications leading to Masters in Science (MS), public health (MPH), and PhD. These courses offer some credits to the eligibility for writing the certification examinations.

The undergraduate medical and nursing curriculum includes a limited introduction to medical genetics and genomics. In brief, there is tremendous enthusiasm and growing appetite among physicians to enhance knowledge and skills to practice genomic medicine. The review by Rubanovich [1] summarizes the key elements for genomic medicine education: identifying and studying ways to best implement low-cost dissemination of genomic information; emphasizing genomic medicine education program evaluation; and incorporating interprofessional and interdisciplinary initiatives.Genomic medicine education and training will become more and more relevant in the years to come as physicians increasingly interact with genomic and other precision medicine technologies.

4.3 The Latin America and the Caribbean

Most nations of the Latin America and the Caribbean continent have provision for medical and human genetic courses at graduate and postgraduate levels. Several universities and major academic organizations are equipped for research that regularly train scientists with PhD or similar degree courses.

Latin America and the Caribbean region make up one of the largest areas of the world, and this region is characterized by a complex mixture of ethnic groups sharing Iberian languages. The area is comprised of nations and regions with different levels of social development. This region has experienced historical advances in the last decades to increase the minimal standards of quality of life; however, several factors, such as concentrated populations in large urban centers and isolated and poor communities, still have an important impact on medical services, particularly genetics services. Latin American researchers have greatly contributed to the development of human genetics, and historic inter-ethnic diversity and the multiplicity of geographic areas are unique for the study of gene-environment interactions. As a result of regional developments in the fields of human and medical genetics, the Latin American Network of Human Genetics (Red Latinoamericana de Genética Humana—RELAGH) was created in 2001 to foster the networking of national associations and societies dedicated to these scientific disciplines. RELAGH has developed important educational activities, such as the Latin American School of Human and Medical Genetics (ELAG), and has held three biannual meetings to encourage international research cooperation among the member countries and international organizations. Since its foundation, RELAGH has been admitted as a full regional member to the International Federation of Human Genetics Societies [2].

The RELAGH is a virtual society of Human Genetics. Nowadays, it represents Latin America at the International Federation of Human Genetics Societies (IFGHS). It is building a set of tools to enhance cooperation and integration of persons, groups, and societies in the continent, a growing need since the Latin American Association of Genetics (ALAG) disbanded its operations in the mid-1990s. In cooperation with the "Genetics for All Institute" (*Instituto Genética para Todos*—IGPT), it organizes the Latin American School of Human and medical Genetics (ELAG). It is a popular annual course that aims to promote a "state-of-the-art" course in Medical Genetics, exposing during 1 week a selected number of top-grade Latin American students with the first-rank researchers in the field, aiming to create new links among Latin American groups [3].

4.4 Indian subcontinent and South Asia

4.4.1 *Medical genetics and genomics*

Graduate and postgraduate courses

Several Indian universities and affiliated colleges offer graduate (BSc-Hon) and postgraduate (MSc) courses in applied biology, human biology, genetics, human genetics, biotechnology, and other related life sciences. In most cases, these provide a sound base for undertaking teaching, pharmaceutical, and related industrial jobs and higher research leading to PhD. With the rapid expansion and new opportunities in diagnostic genomics, a young scientist, equipped with a Masters or preferably PhD in Human Genetics or a related discipline, would be a highly sought after by these emerging genomic laboratories. Few young science graduates, who opted to study medicine, find these courses very helpful in their pursuit for a career in medical genetics and genomics.

DM in medical genetics

Few selected postgraduate Indian medical colleges and institutes offer higher specialist training in medical genetics leading to the Doctorate in Medicine (DM) degree. It is regulated by statute of the National Medical Commission of India (www.nms.org.in). This is a three-year competency-based postgraduate training program in medical genetics. Typically, a doctor holds a postgraduate degree (MD) in medicine, pediatrics, and obstetrics and gynecology.

The competency-based training for DM in Medical Genetics aims to produce a postgraduate student who after undergoing the required training should be able to deal effectively with the needs of the patients, and community, and should be competent to handle medical problems related to genetic disorders. These include clinical evaluation, investigations, genetic workups requiring pre-test and post-test counseling, up-to-date information, and abilities to carry out novel treatments and skills for planning and implementation of population-based prevention programs. Last but not least, be ready for carrying out clinical practice of personalized medicine in the 21st-century molecular medicine era. The postgraduate student should also acquire skills to teach medical genetics to undergraduates and paramedical students as well.

The objectives of the DM in Medical Genetics course are to produce a competent and skilled medical geneticist who is:

- a medical doctor who can evaluate a patient with possible genetic disorder, ascertain the risk of a genetic disorders, make a clinical diagnosis, able to decide appropriate test to confirm the diagnosis, and provide the latest form of treatment. The other goal is genetic counseling to the patient, family, and extended family about management, carrier screening, prenatal diagnosis, and prevention.
- aware of contemporary advances and developments in Medical Genetics, and ways for continued learning for keeping updated about diagnostic investigations, technological developments, and treatment.
- trained in use of novel clinical ways of phenotyping, use of software for phenotypingand correlation with genotypes, interpretation of genetic variants and CNVs regarding their pathogenicity and causal nature and comfortable with the use of databases of genetic disorders, CNVs, and genetic variants.
- competent pertaining to medical genetics that is required for practice in the community and at all levels of health system, especially population-based prevention program, screening for late-onset disorder, and susceptibility to cancer.
- informed of the health needs of patients and families with genetic disorders and carries out professional obligations in keeping with principles of the National Health Policy and professional ethics.
- able to identify the disorders that are prevalent in Indian populations (identifying pockets).
- oriented to principles of research methodology.
- skilled in educating lay persons, medical and paramedical professionals.
- skilled in effectively communicating with the person, family, and community.
- skilled in DNA diagnostic tests, including those for screening of carriers and antenatal diagnosis, as well as premorbid diagnosis for primary prevention (predictive/preventive medicine), being aware of the PCPNDT 2003 Act and related guidelines as well as process.
- competent with knowledge and skills in genetic counseling with required confidence in dealing with psychological and social issues.

The syllabus and training schedule are comprehensive for DM in Medical Genetics. Assessment and evaluation process is systematic and robust. There are now several leading medical genetic centers in academia, tertiary healthcare provider hospitals, and dedicated medical genomic laboratories across India that employ highly qualified doctor with DM in Medical Genetics.

DNB in medical genetics

In 1975, the Government of India established the National Board of Examinations (NBE), similar to the American Board Examinations. The Diploma of NBE (DNB) Program in Medical Genetics is designed to give physicians a thorough knowledge of the principles and practice of medical genetics, and prepare them for a leadership role in training other medical geneticists for service and research.

The candidates are provided hands-on training in modern genetic technologies such as polymerase chain reaction (PCR), Sanger sequencing, massively parallel sequencing, microarrays, enzyme assays, FISH, and Luminex multiplexing. They are imparted with knowledge of the principles of epidemiology and statistics as applied to genetic and genomic research. They are given the opportunity to carry out research on a chosen topic. The curriculum includes the ethical principles as applied to genetic services and practice.

The main goals of the National DNB Program in Medical Genetics are:

(i) Impart training that will enable the trainees to evaluate patients with genetic disease, order appropriate tests, interpret them for the patient, make a precise diagnosis, and provide genetic counseling.

(ii) Inform about screening pregnant women for genetic disease, and take appropriate action to prevent birth of children with malformations and genetic disorders.

(iii) Provide skills in laboratory genetics to enable them to establish genetic tests using chromosomal studies, biochemical assays, and molecular techniques.

(iv) Ensure that the students acquire necessary knowledge and skills to plan and carry out research and perform statistical analysis of the data generated.

(v) Acquaint candidates with the principles of ethics as applied to genetic services and research.

(vi) Train the candidates to think independently and become leaders in setting up genetic services and carrying out research.

List of training programs in Medical Genetics and Genetic Counseling in India.

Degree	Eligibility	Duration	Centers
DM (Medical Genetics)	Basic medical qualification recognized by the Medical Council of India (i.e., MBBS or an equivalent degree) and a postgraduate medical degree in Pediatrics/Internal Medicine/Obstetrics and Gynecology recognized by the Medical Council of India (i.e., MD/MS/DNB or an equivalent degree)	3 years	Sanjay Gandhi Postgraduate Institute of Medical Sciences, Lucknow All India Institutes of Medical Sciences, New Delhi
DNB Medical Genetics	MD or DNB in Pediatrics/Medicine/Obstetrics and Gynecology or equivalent degree recognized by the MCI	3 years	National Institute of Biomedical Genomics, West Bengal Nizam's Institute of Medical Sciences, Hyderabad Sir Ganga Ram Institute of Postgraduate Medicine and Research, New Delhi

Continued

List of training programs in Medical Genetics and Genetic Counseling in India—cont'd

Degree	Eligibility	Duration	Centers
M.Sc. Biomedical Genetics/ Biomedical Genetics with specialization in Genetic Counseling	B.Sc. degree from a recognized university in any branch of Biological Sciences that includes Botany/Zoology, Microbiology/ Biochemistry/Biotechnology/Nutrition (or) Home Science/Agricultural Science/ Medicine with 50% of marks (inclusive of all subjects)	2 Years	Vellore Institute of Technology (VIT), Vellore
M.Sc. Genetic Counseling	MBBS /BDS /BSc Nursing /Human Genetics /Human Biology /Medical Biotechnology /Molecular Biology, and Allied Health from a recognized university	2 years	Kasturba Medical College, Manipal
	B.Sc. Life Sciences preferably degree in Nursing or Human Genetics or Human Biology or Medical Biotechnology or Molecular Biology with aggregate percentage of 55% and above OR MBBS degree recognized by Medical Council of India	2 years	Nizam's Institute of Medical Sciences, Hyderabad
SIAMG-GENZYME Clinical Genetics fellowship	Basic medical qualification recognized by the Medical Council of India (i.e., MBBS or an equivalent degree) and a postgraduate medical degree in Pediatrics/Internal Medicine/Obstetrics and Gynecology/Dermatology/ Ophthalmology/ Radiology/Surgery/Orthopedics recognized by the Medical Council of India (i.e., MD/MS/DNB or an equivalent degree)	3 months	Sanjay Gandhi Postgraduate Institute of Medical Sciences, Lucknow All India Institutes of Medical Sciences, New Delhi Christian Medical College and Hospital, Vellore Kasturba Medical College, Manipal Sir Ganga Ram Institute of Postgraduate Medicine and Research, New Delhi Nizam's Institute of Medical Sciences, Hyderabad Sri Avittom Tirunal Hospital Medical College, Thiruvananthapuram

Source Indian Academy of Medical Genetics www.iamg.in.

Fellowship in clinical genetics

Several universities, academic institutions, and hospitals across India have introduced a number of fellowship training opportunities for doctors and laboratory scientists. These are primarily adjunct to formal medical genetic training and degree courses as discussed in previous sections. Typically, a doctor with fellowship in clinical or medical genetics would be expected to have basic understanding of genetics, grasp on core principles of genetic counseling and specific information on key genetic disorders in the given area or specialty. Most of these doctors and related healthcare practitioners practice independently or affiliated with a hospital. There are currently several such medical practitioners around who offer specialist service in reproductive genetics and fetal medicine, inherited metabolic diseases, pediatric and adult neurology, cardiovascular medicine, rheumatology, orthopedic surgery, and oncology. Most of these practitioners are not acknowledged or registrable with either the respective State Medical or Health Council.

Goals and objectives of the fellowship course in pediatric genetics offered by one of the universities (www.igich.insets).

Goals

- Construct and interpret a three-generation pedigree in a family with suspected genetic disorder
- Evaluate an infant or a child with a common genetic or inherited metabolic disorder
- Arrange appropriate genetic tests with locally available resources
- Interpret the results of laboratory genetic investigations, including chromosomes, FISH, and DNA tests
- Counsel families with the commonest genetic disorders presenting in pediatric age group
- Organize family screening of other members at risk, including appropriate prenatal diagnosis in a subsequent pregnancy
- Appropriately refer to senior colleagues where basic workup has been completed

Objectives

- **Knowledge**—to equip with ability to recognize, investigate, manage, and counsel the commonest genetic disorders seen in pediatric practice. It also includes mandatory rotation in genetic laboratory, interpretation of genetic diagnostic results, in-depth family history, risk assessment and decision-making skills for prenatal diagnosis, multidisciplinary team working and participation in therapeutic interventions.
- **Skills**—to acquire appropriate skills in the specialty for providing quality care to children and adolescents needing genetic evaluation and to provide counseling to their parents in order to help them attain the optimal results in the management of their child's disorder. The course will also provide trainees with skills of evaluation a pedigree, assessing genetic risk, interpreting genetic test results, and the ability to professionally offer genetic counseling to these families.
  Communication abilities—Genetic counseling is an important and integral part of the management of a genetic disorder. In a country where the majority of individuals are unfamiliar with the basic concepts of genetics and inheritance, the ability to make a family understand the basis of the genetic disorder in their family and its implications requires skill and good communication abilities.

4.4.2 Genetic counseling

The status of genetic counseling across the Indian subcontinent and south Asia is similar to early years in most present-day developed Western nations. Currently, separate Board of Genetic Counseling or equivalent regulatory bodies exist in USA, Canada, UK and Europe, Australasia, and South Africa. In most places, the knowledge component is structured in the curriculum of Master's courses (MSc or MS). Skills and competencies components are expected to be completed through active participation in supervised genetic clinics. Each genetic counselor trainee is required to develop a classified and validated cases portfolio. The supervisor is expected to ensure that learning objectives were achieved with a satisfactory evaluation.

During the last 20 years, several genetic counseling (GC) courses have emerged across India. Most are knowledge based with either MSc or Diploma postgraduate qualifications. In addition, there are few certification courses. Apart from few places, there is no opportunity or requirement to undergo clinical training; thus, most fresh genetic counselors lack requisite skills and competencies. This gap is often noticed in actual clinical settings. The Board of Genetic Counseling in India, a nongovernmental not-for-profit organization, has developed a structured knowledge and skills competency-based genetic counseling training program (http://www.geneticcounselingboardindia.com). The Board undertakes several roles and responsibilities. These include develop GC training programs by creating a common curriculum, accredit the current and new GC programs, offer GC Certification to for practice in India, set rules to provide certification to existing genetic who currently provide GC services, and fully meet established criteria for standards and clinical experience, ensure that hospitals, nursing homes, clinics, government agencies, biotechnology companies, diagnostic centers, or other professional settings employ or have access to a certified professional GC for providing appropriate healthcare, and certify GCs to enable them to work as independent healthcare professionals in India and other developing nations.

The Indian Board of Genetic Counseling is actively pursuing official recognition and status of genetic counseling as a recognized profession with defined skills and competencies. The Board has made representation with evidence to the National Skill Development Corporation, Ministry of Skill Development and Entrepreneurship (www.nsdcindia.org). The evidence included recommendations from the Association of Genetic Nurses and Counselors (AGNC), a constituent organization within the British Society of Genetic Medicine (www.bsgm.org.uk). This approach has led to the inclusion of genetic counseling as a distinct occupation with defined skills and competencies by the healthcare sector skill council (www.healthcare-ssc.in).

4.4.3 Laboratory genetics

There are several academic and industry-based genetic and genomic laboratories. Apart from occasional specific clinical input, these units do not engage in any diagnostic or therapeutic genetic or genomic investigations. However, these offer Master's and PhD programs in different branches of applied biology, cell and molecular biology, biotechnology, genetic engineering, and computational biology.

During the last 15 years, new private genetic and genomic laboratories have established in different parts of South Asia. These offer a range of diagnostic services including the microarray chromosome analysis, single and multigene panel DNA analysis, next-generation small or large gene panel testing, clinical or whole-exome sequencing, and whole-genome sequencing. Few laboratories also offer RNA sequencing and proteomic analysis. These laboratories offer non-accredited in-house training with the sole purpose of developing specific skills and competencies. The public-funded diagnostic genetic laboratory (Centre for DNA Fingerprinting and Diagnostics, CDFD, www.cdfd.org.in) in Hyderabad (State of Telangana in south India) offers a range of diagnostic genetic testing and training opportunities for career laboratory scientists in genetics and genomics.

4.4.4 Genetic and genomic research

Considerable opportunities for training and skills development in India also exist through genetic and genomic research funded by the Indian Council of Medical Research (www.icmr.gov.in), the Council of Scientific and Industrial Research (CSIR; www.csir.res.in), and Department of Biotechnology (www.dbtindia.gov.in). Contributions made by the dedicated Institute of Genomics and Integrated Biology (IGIB; www.igib.res.in) and the Wellcome-DBT India Alliance (www.indialliance.org) are acknowledged by young and emerging science graduates and postdoctoral fellows and as well as junior, middle cadre, and senior career scientists. These institutions have helped to enhance skills and competencies among many professionals working in basic and applied genetics and genomics.

The Middle East and Arab Nations

Since the first publication of the Late Professor Teebi's multiauthor edited reference book on genetic disorders of the Arab world (Ref…), the awareness and preparedness for dealing genetic diseases across the Arab nations have considerably grown (Ref. Al-Gazali; Bayoumi). Massive investments in developing the large-scale population-based genome sequencing database have led to dedicated resources such as DALIA (Vatsayan et al.…) and the Qatar Genome Programme (www.qatargenome.org.qa; [4]).

There are practicing clinical geneticists, other clinicians with special interest in genetics, and genetic counselors spread across the Middle East and the Arab world. Most of these received their informal and formal structured teaching and training in either North America or Europe. Few were educated and trained in India, South Africa, and Australasia. Apart from short courses and seminars, there are no courses or seminars in the Arab world dedicated to genetic or genomic education and training.

4.6 China, Japan, and Asia Pacific

China, Japan, and small nations in the far eastern and several islandic countries in the Asia-Pacific appeared late on the global medical genetic horizon. However, rapid progress has been made since the completion of the Human Genome Project. Interest in genetic diseases and traits in China dates back to 1937 when the frequency of X-linked color blindness was analyzed. Since then, many genetic diseases, such as G6PD deficiency, Down syndrome, and abnormal hemoglobin diseases, have been studied. In 1962, the Medical Geneticist Min Wu established the division of Human Cytogenetics at the Department of Pathology in the Institute of Experimental Medicine, Chinese Academy of Medical Sciences [5]. Since then, several academic institutions have established disease-oriented medical genetic units in different parts of China. This development led to some recognition by the Chinese Academy of Sciences. First Division of Medical Genetics was established at the Internal Medicine Department in Peking Union Medical College Hospital. Subsequently, more and more domestic doctors and geneticists became involved. The first clinic to offer genetic counseling services was set up in 1972 at the Xiangya Hospital in Central South University.

In 1978, the Chinese government implemented the reform and opening-up policy encouraging further development of medical genetics. This led to the founding of the National

Committee of Human and Medical Genetics under the leadership of the Chinese Society of Genetics in 1979. This committee consisted of eight specialty groups among which groups of internal genetic medicine, pediatric genetic medicine, neuropsychiatric genetic medicine, and prenatal diagnostic medicine were closely related to clinical genetics [6]. Seven years later, the medical section of the committee formed an independent society, the Association of Medical Genetics, under the leadership of the Chinese Medical Association. During this period, several medical colleges, including Sun Yat-sen Medical University, opened new departments to teach medical genetics. However, formal structured and regulated medical genetic training programs for doctors are not available. Most doctors developed the medical genetic knowledge base with specific clinical skills, such as prenatal diagnosis, on an individual basis.

The genetic counseling as a distinct profession emerged in China relatively late but has rapidly grown over the last few years. Prospects of genetic counseling as a profession look good. Genetic counselors play a pivotal role in building communication channels between medical doctors and their patients, the government and the general public, and social organizations and their customers in China. The growth of genetic counseling aims to enable patients and family members to make informed decision, which in turn will lead to the reduction of the birth prevalence of severe congenital anomalies and genetic disorders.

In February 2015, the Chinese Board of Genetic Counseling (CBGC) was founded (http://www.cbgc.org.cn). Within 2 months, the CBGC launched its first training course in Shanghai introducing basic genetic counseling concepts. This course perfectly aligned with the expansion of noninvasive prenatal testing (NIPT) at hospitals and attracted hundreds of hospital directors and medical doctors. The CBGC continues to be actively engaged in genetic counseling teaching and training in the mainland China (Fig. 3). The status of medical genetics and genetic counseling in Hong Kong and Taiwan is comparatively much developed

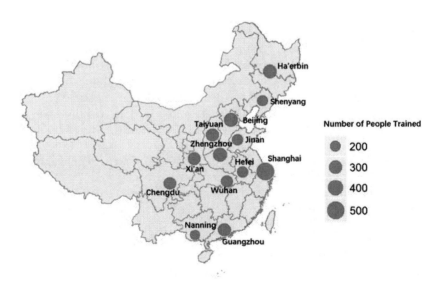

FIG. 3   State-wide number of people trained by the Chinese Board of Genetic Counselors until April 2018. *Adopted with permission from as the American Journal of Medical Genetics. Part C, Seminars in Medical Genetics, 181(12) (2019) 170–176.*

and established. There are formal teaching and training programs for both sectors funded by the Department of Health (www.dh.gov.hk) and the University of Hong Kong (www.med.khu.hk). In Taiwan, the State-funded Institute of Genomic Medicine is pivotal for teaching, training, research, and development in basic and applied aspects of genomics (www.cgm.ntu.edu.tw). There is an active program to introduce and encourage the practice of genomic medicine among physicians [7].

In Japan, the Japanese Society of Genetic Counseling, founded in 2001, currently has more than 100 genetic counselors (http://www.jsgc.jp). The genetic counseling profession in other Asian countries, such as Korea, Singapore, Thailand, and Malaysia, is still in its infant stages. While the development of the medical genetics and genetic counseling profession in the far east is relatively preliminary, recent strong support from the government and concerted efforts of overseas experts and domestic geneticists and clinicians have inspired breakthroughs in the field of clinical genetics, genetic counseling, and genomic medicine.

4.7 Australasia

The status of medical genetics and genomics in Australia and New Zealand is probably similar to the Western developed nations of Europe and North America. However, there might be some variations. Most tertiary care hospitals, usually state-funded, host and offer a wide range of clinical genetic and genetic counseling services. There are structured and state-regulated teaching and training programs for doctors, nurses, counselors, and other therapists who take up clinical and academic positions. The curriculum and assessment criteria are robust and stringent (see Chapter 2). Most Australasian universities offer Master's and higher Doctoral training in basic and applied aspects of genetics, genomics, and other *omic* sciences.

The Australian Government has committed and generously funded the Australian Genome Project (www.australiangenomics.org.au; [8]). It is a national multicenter collaborative research partnership of more than 80 organizations piloting integrating genomics into medicine and healthcare. The primary is to assess the application of genomic testing in healthcare with targeted research and development. Emphasis is on clinical projects in rare diseases, cancers, and reproductive carrier screening. In addition, there are programs for advancing the diagnostic, health informatics, regulatory, ethical, policy, and workforce infrastructure necessary for the integration of genomics into the Australian health system. The project is managed and administered through the Australian Genomics Health Alliance (www.melbournebioinformatics.org.au). Around 80 organizations participate in this collaborative research partnership (Fig. 4; [8]).

5 Conclusion

Globally, phenomenal efforts and achievements are made in teaching and training medical and healthcare professionals and scientists to acquire new knowledge and develop skills and competencies considered essential for managing diverse genetic or inherited conditions. Apart from highly resourced Western nations, few less developed and emerging developing nations have acknowledged the need for skilled and competent workforce. There is clear public and political will to fill this gap.

FIG. 4 Australian Genomics: A collaborative research partnership of more than 80 organizations. Z. Stark, T. Boughtwood, P. Phillips, J. Christodoulou, D.P. Hansen, J. Braithwaite, A.J. Newson, C.L. Gaff, A.H. Sinclair, K.N. North, Australian genomics: a federated model for integrating genomics into healthcare, Am. J. Hum. Genet. 105 (1) (2019) 7–14; adopted with permission from American Journal of Human Genetics.

Acknowledgments and disclaimer

The author acknowledges input and support offered by several colleagues in searching the literature and putting together factual and applied information. A reader will find data and information obtained from few public online domains. The author does not make any claim on the ownership and/or expertise in the information included in this chapter.

References

[1] C.K. Rubanovich, C. Cheung, J. Mandel, C.S. Bloss, Physician preparedness for big genomic data: a review of genomic medicine education initiatives in the United States, Hum. Mol. Genet. 27 (R2) (2018) R250–R258, https://doi.org/10.1093/hmg/ddy170.

[2] A. Rojas-Martínez, A. Giraldo-Ríos, G. Jiménez-Arce, A.F. de Vargas, R. Giugliani, RELAGH—the challenge of having a scientific network in Latin America: an account from the presidents, Genet. Mol. Biol. 37 (1 suppl. 1) (2014), https://doi.org/10.1590/S1415-47572014000200017.

[3] R. Giugliani, G. Baldo, U. Matte, Latin American School of Human and Medical Genetics and Latin American Network of Human Genetics: promoting education, interaction and networking across Latin America, Chapter, in: D. Kumar (Ed.), Genomics and Health in the Developing World, Oxford University Press, NY, 2012, https://doi.org/10.1093/med/9780195374759.001.0001. 10.1093/med/9780195374759.003.0107.

[4] G. Thareja, Y. Al-Sarraj, A. Belkadi, M. Almotawa, The Qatar Genome Program Research (QGPR) Consortium, K. Suhre, O.M.E. Albagha, Whole genome sequencing in the Middle Eastern Qatari population identifies genetic associations with 45 clinically relevant traits, Nat. Commun. 12 (2021), 1250.

[5] L. Sun, B. Liang, L. Zhu, Y. Shen, L. He, The rise of the genetic counseling profession in China, Am. J. Med. Genet. C Semin. Med. Genet. 181 (2) (2019) 170–176, https://doi.org/10.1002/ajmg.c.31693. PMC6593421 30860676.

[6] C.G. Li, Development situation and prospects of medical genetics in China, in: Paper Presented at the China Healthy Birth Science Academic Exchange Conference in 2004. Chengdu, Sichuan Province, China, 2004.

[7] L.-S. Chen, F.-W. Chang, M. Kim, D. Talwar, S. Zhao, Genomic medicine practice among physicians in Taiwan, Pers. Med. 14 (2) (2017), https://doi.org/10.2217/pme-2016-0067. RESEARCH ARTICLE, Published Online: 24 Jan 2017.

[8] Z. Stark, T. Boughtwood, P. Phillips, J. Christodoulou, D.P. Hansen, J. Braithwaite, A.J. Newson, C.L. Gaff, A.H. Sinclair, K.N. North, Australian genomics: a federated model for integrating genomics into healthcare, Am. J. Hum. Genet. 105 (1) (2019) 7–14.

Preparing the workforce for genomic medicine: International challenges and strategies

Desalyn L. Johnson[a], Bruce R. Korf[a], Marta Ascurra[b], Ghada El-Kamah[c], Karen Fieggen[d], Beatriz de la Fuente[e], Saqib Mahmood[f], Augusto Rojas-Martinez[g], Ximena Montenegro-Garreaud[h], Angelica Moresco[i], Helen Mountain[j], Nicholas Pachter[k,l,m], Ratna Dua Puri[n], Victor Raggio[o], Nilam Thakur[p], and Rosa Pardo Vargas[q]

[a]The UAB Heersink School of Medicine, Birmingham, AL, United States [b]National Program for the Prevention of Congenital Defects-Ministry of Public Health and Social Welfare of Paraguay, Asunción, Paraguay [c]Human Genetics and Genome Research Institute, National Research Centre, Cairo, Egypt [d]University of Cape Town, Cape Town, South Africa [e]Department of Genetics, Faculty of Medicine and Dr. Jose Eleuterio Gonzalez University Hospital, Autonomous University of Nuevo Leon, San Nicolás de los Garza, México [f]Department of Human Genetics & Molecular Biology, University of Health Sciences, Lahore, Pakistan [g]School of Medicine and Health Sciences, Tecnologico de Monterrey, Monterrey, Mexico [h]Institute of Medical Genetics, Lima, Perú [i]Department of Genetics, J. P. Garrahan National Pediatric Hospital, Buenos Aires, Argentina [j]Familial Cancer Program, Genetic Services of Western Australia, Subiaco, WA, Australia [k]Genetic Services of Western Australia, King Edward Memorial Hospital, Perth, WA, Australia [l]School of Medicine and Pharmacology, University of Western Australia, Perth, WA, Australia [m]School of Medicine, Curtin University, Perth, WA, Australia [n]Institute of Medical Genetics & Genomics, Sir Ganga Ram Hospital, New Delhi, India [o]Genetics Department, Faculty of Medicine, University of the Republic, Montevideo, Uruguay [p]National Academy of Medical Sciences, Bir hospital, Kathmandu, Nepal [q]University of Chile Clinical Hospital, Santiago, RM, Chile*

1 Introduction

Implementation of genomic medicine will require awareness, education, and user-friendly systems for all health providers, as well as a group of specially trained providers who are experts in genomic medicine. Genomic medicine is a relatively new area of medicine, so there are few established paradigms of training in the area. Furthermore, there are currently large disparities in access to expertise in genomic medicine around the world. In this chapter, we will consider the needs for a genomic medicine workforce and some strategies for this to be developed, particularly in middle and lower-income countries globally.

2 The genetic architecture of disease and scope of practice in genomic medicine

Genetic factors contribute to virtually all aspects of health and disease, though the genetic architecture of disease can range widely [1]. At one extreme, changes in just a single gene, or gross changes in chromosomal structure that affect multiple genes, can determine a specific phenotype. Examples of single gene disorders include cystic fibrosis, sickle cell disease, and neurofibromatosis (types 1 & 2); chromosomal abnormalities can include trisomy (e.g., trisomy 21, Down syndrome) or monosomy (monosomy X, Turner syndrome) involving an entire chromosome or copy number variation (usually deletion or duplication) of a specific chromosome region (e.g., 22q11.2 deletion syndrome). At the opposite extreme are disorders resulting from an interaction of multiple genes and environmental factors, referred to as multifactorial inheritance. These conditions include most instances of common disorders such as hypertension, diabetes (type 1 or type 2), and asthma. Until recently, the specific genes that contribute to common disorders were mostly unknown, but genome-wide association studies have revealed many genes associated with common disorders [2]. Polygenic risk scores can be calculated to identify individuals at high risk of disease who may benefit from risk-reducing interventions [3], though the preponderance of data currently available was obtained from European ancestry populations [4], raising the question of portability to other ancestries.

Clinical geneticists have traditionally focused their practice on the prevention, diagnosis, and treatment of single gene and chromosomal disorders. They provide risk assessment, ordering and interpreting diagnostic tests, and, in some cases, management programs consisting of surveillance and possibly treatment. Laboratory geneticists provide diagnostic testing, including both genetic and genomic testing, as well as biochemical genetic testing. As the power of genomic analysis increases, the boundaries of the scope of practice in medical genetics are becoming blurred. Genetic testing is increasingly available for the diagnosis of single gene conditions that may affect just a single body system, for example, cardiomyopathy or neurodegenerative disease. Genetic diagnosis may be performed entirely by specialists such as cardiologists or neurologists, or, in some instances, physician geneticists and/or genetic counselors may assist. The role of the clinical geneticist in these cases may depend on how the workforce is structured in a region and on the specific patterns of practice in each medical institution.

In this chapter, we will define the term "genomic medicine" as the use of genomic data to inform medical decisions about the prevention, diagnosis, and treatment of disease. It is likely that genomic testing for the determination of polygenic risk scores or pharmacogenetic variants will gradually be incorporated into medical practice and be ordered and interpreted by a wide range of medical practitioners. Clinical geneticists may be involved in the interpretation of complex cases and genetic counselors, and nurses may work with physicians in different specialties, but it is unlikely that all genetic tests will be ordered by clinical geneticists in the future. This is already well established in the management of cancer, where oncologists order genome sequencing of tumors to guide treatment. Occasionally, germ line variants may be detected that have implications for other family members and might be managed by a physician geneticist or genetic counselor, but the interpretation of the tumor sequencing itself is usually done by a molecular pathologist providing a report to an oncologist.

3 Types of medical genetics professionals

Medical genetics professionals include clinical and laboratory geneticists. Although the specific types of practitioners available in different regions vary, some of the major categories are defined below. We will describe the training provided for these specialists in the United States [5], and briefly note the global status of training.

3.1 Physician clinical geneticists

Physician clinical geneticists provide risk assessment, order, and interpret genetic tests, institute medical surveillance for individuals with genetic disorders, and prescribe indicated treatments. Some may specialize in specific areas, such as perinatal genetics, biochemical genetics, pediatric or adult genetics, neurogenetics, cancer genetics, etc., whereas others may practice more broadly. In the US, physicians training in Clinical Genetics and Genomics must complete at least one year of training in an American Council of Graduate Medical Education (ACGME)-accredited medical residency (e.g., internal medicine, pediatrics, etc.) and then two years of specialty training in a medical genetics residency accredited by the ACGME. Four-year training programs are offered combining medical genetics with internal medicine, pediatrics, maternal-fetal medicine, or reproductive endocrinology and infertility. Candidates must pass an examination given by the American Board of Medical Genetics and Genomics (ABMGG) to achieve board certification; after board certification, practitioners must participate in a maintenance of certification (MOC) program. Subspecialty training is offered in Medical Biochemical Genetics, requiring an additional year of training in biochemical genetics for medical geneticists or two years of biochemical genetics training for physicians certified in another medical specialty. Formal specialty training for physician clinical geneticists is also offered in other high-income countries, including Canada, the United Kingdom, countries of the European Union, and Australia.

3.2 Genetic counselors

Genetic counselors provide assessment and counseling to patients and families regarding genetic risks, results of genetic tests, and options for managing risk of disease or an established diagnosis. In the US, genetic counselors are certified by the American Board of Genetic

Counseling. Training involves completion of a master's degree program (usually two-year duration) accredited by the Accreditation Council for Genetic Counseling. Training includes didactic and clinical training in genetics and in the principles of counseling. Diplomates must recertify every five years either by taking an examination or by participating in specific continuing education activities. Genetic counseling is not as recognized as a formal profession as widely internationally as physician geneticists, but formal training programs are increasing. The worldwide status of genetic counseling was reviewed by Abacan et al. in 2019 [6]. At that time, they estimated that there were upwards of 7000 genetic counselors in at least 28 countries, though the numbers have likely increased since that survey.

3.3 Nurse geneticists

In some areas, nurses may obtain additional training in medical genetics. Depending on their level of training, they may work in collaboration with a physician providing diagnosis, management, and treatment of individuals with genetic disorders.

3.4 Specialty genetics training in middle- and low-income countries

Formal training and recognition of various types of medical genetics practitioners is less well established in middle- and low-income countries. Some general information about the status of genetics training in these regions is provided below.

3.4.1 Latin America

Standardized genetics education requirements are not present in all Latin American countries. However, international organizations such as RELAGH (Latin-American Network of Human Genetics) [7] have pursued initiatives to standardize genetics education within the continent. Countries with more developed programs include Argentina [8], Brazil [9], Chile, Cuba, Colombia, Ecuador [10], Mexico, and Peru [11]. These countries all have acknowledged medical genetics as an official specialty, which usually requires three years of residency training after medical school. In addition, Mexico has subspeciality programs in neonatal screening, neurological and ophthalmological disorders, among others. Cuba is the only Latin country that has formal training for genetic counselors; plans are underway to establish a program in Brazil. In most other Latin American countries, the responsibility of genetic counseling remains with the physician. Across the continent, there is a variety in the available training for cytogenetics, molecular and biochemical genetics.

3.4.2 South Asia

There are few formal genetics training programs in the region. Nepal and Pakistan have no genetics education program; trainees must travel abroad to receive training. India has a pathway to becoming a clinical geneticist [12]. This consists of obtaining a bachelor's in medicine (4.5 years), followed by a master's degree in either pediatrics, OB/GYN, or internal medicine. Once the 3-year master's degree is completed, a student can enter a 3-year Genetics Residency Program. In addition, a master's program in Genetic Counseling has recently been established. National organizations such as the Indian Society of Human Genetics and the Indian Academy of Medical Genetics have been established to enhance genetics education and services. The Indian Academy of Medical Genetics has a quarterly publication, *Genetic Clinics*, which discusses recent advances in the field. Additionally, the members of the academy participate in monthly tele-meetings where interesting cases are discussed. The *Indian Journal of Human Genetics* is a publication established to increase genetics research in India.

3.4.3 East Asia

Several countries in East Asia have established training programs for clinical geneticists, including South Korea, Thailand, the Philippines, Malaysia, and Hong Kong. These training programs typically consist of 2–3 years of clinical genetics training after medical school. In Singapore, public hospitals such as the National University Hospital and KK Women's and Children's Hospital provide subspecialty fellowship training. However, the Singapore Medical Council and Subspecialty Accreditation do not recognize clinical genetics as an official subspecialty. Although Indonesia does not have a formal clinical genetics residency program, its first Human Genetics Association was established in 1976. After years of inactivity, it was reborn as the Indonesian Society of Human Genetics (InaSHG) in 2016. A major goal of InaSHG is for the Indonesian Medical Doctor Association and the Ministry of Health of Indonesia to recognize and license Genetics professionals. Hong Kong has the youngest clinical geneticist training program in the region, which began accepting applicants in January of 2017. South Korea has extensive requirements for obtaining and maintaining status as a clinical geneticist. These requirements include one year of practical clinical genetics experience (150 cases), attendance at yearly conferences, membership in the Korean Medical Genetics Society, a publication in the *Journal of Genetic Medicine*, and a qualification examination every 5 years.

Several countries have established genetic counseling programs, including the Philippines, Taiwan, Indonesia, and Malaysia. Although Indonesia has no formal clinical genetics training program, it created one of the first Masters of Genetic Counseling programs in Asia. Taiwan has an official master's degree program for genetic counseling; however, there is no formal national examination or certification. Malaysia has an established Masters of Genetic Counseling program; however, they are not officially recognized by the Counseling Board of Malaysia. The Genetic Counseling Society of Malaysia was founded in 2018 and seeks to obtain official recognition of their profession.

There is a scarcity of programs designed to specifically provide training in cytogenetics, molecular, and biochemical genetics.

3.4.4 Africa

Formal programs for training medical genetics professionals exist in Egypt and South Africa. In Egypt, the requirements for clinical geneticist include six years of medical school, followed by a 3-year residency to receive a master's degree in genetics, pediatrics, internal medicine, or phoniatrics followed by a PhD or MD in the same previous specialties. In South Africa, the training for clinical geneticists is slightly different. Medical school (six years) is followed by two years on medical internship and then one year as a community service medical officer prior to registration for independent practice. This is followed by four years of specialty training in medical genetics. A two-part national fellowship examination consisting of written papers, portfolio assessment, and oral/clinical examination is also required.

In Egypt, cytogenetics, molecular genetics, and biochemical genetics all follow a similar pathway: a bachelor's in medicine and surgery, science, or pharmacy; followed by a master's degree and a MD or PhD. In South Africa, cytogenetics requires a bachelor's degree and laboratory work. Molecular genetics also requires a bachelor's degree, but most scientists in the field have at least a master's degree. To sign diagnostic reports, scientists must register with the Health Profession Council of South Africa. This process requires an internship and portfolio assessment. No formal program for biochemical genetics is established in South Africa.

In Egypt, there is no official genetic counseling program; this service is offered by clinical geneticists. In South Africa, genetic counselors must finish a 2-year master's program and two years of internship (with the first year done concurrently with second year of the degree). Counselors must register with the Health Professionals Council of South Africa.

4 Competencies in genomic medicine

The medical genetics workforce will be needed to provide the technical component of genomic testing and need to be available to help interpret results, especially in instances that are not straightforward. Full integration of genomics into medical practice, however, will require health providers who do not have formal training in genomics to play a role. Examples might include using pharmacogenetic information in prescribing a medication or incorporating a polygenic risk score along with other information in managing a patient at risk of coronary artery disease.

In 2014, a group of professionals working with the Intersociety Coordinating Committee for Physician Education in Genomics formulated a framework for physician competencies in genomics [13]. The framework was not intended to stand alone as a list of competencies for every medical specialty; rather, it was intended to provide a foundation from which to build a set of specialty-specific competencies. Genomic medicine has advanced significantly since 2014; for example, polygenic risk scores were in their infancy in terms of development. Although they still may not be ready for routine clinical application, discussion about how to do this and research efforts to test the clinical use of polygenic risk scores are now underway. Table 1 reiterates the competencies as described in 2014, with some additions based on more recent developments in genomic medicine.

The ability to implement these competencies will vary widely around the world, depending on access to genomic testing as well as educational resources and assessment of priorities. It may, however, be more feasible to incorporate genomics into training of health professionals than to identify a subset willing to dedicate their entire professional career to genomic medicine. Several types of initiatives might be considered in enhancing the ability to provide training in genomic medicine in diverse settings. These include:

- Developing local courses and degree programs, such as a master's degree in genomic medicine; a set of best practices for design of genomic education programs has been proposed by Nisselle et al. [24]
- Development of curricular materials that can be utilized in training. Haspel et al. have developed curricular materials initially designed to teach genomics to pathology residents [25].

TABLE 1 Physician competencies in genomic medicine based on ISCC workgroup publication [13], with some updates and comments.

Entrustable professional activity	Comments
Elicit, document, and act on relevant family history pertinent to a patient's clinical status. • Obtain and document family history • Recognize patterns of inheritance • Assess genetic risks • Explain implications of family history to patient/family • Work alongside other professionals, such as genetic counselors • Use major genomic databases (e.g., OMIM, Genetic Test Registry)	Taking a family history remains a challenge in day-to-day practice given the time required. Some approaches have been developed to facilitate this and have been deployed in some limited resource settings [14].
Use genomic testing to guide patient management. • Major approaches to genomic analysis (e.g., panel-based tests, sequencing, copy number variation) • Indications and limitations of genomic tests • Interpretation of test results in light of clinical question • Utilization of genomic databases (e.g., ClinVar) • Use of genomic population screening in newborns and adults • Prenatal and preconceptional testing (including non-invasive prenatal screening from cell-free DNA) • Application of polygenic risk scores • Pre- and post-test counseling of patients/families • Ethical, legal, and social issues in genetic testing	Access to genomic testing will vary widely around the world, as does cost and coverage by health plans. Some specialties, such as maternal-fetal medicine [15], newborn medicine [16], or oncology [17], may be more rapid adopters of new technologies such as analysis of cell-free DNA or genome sequencing. Newborn screening is currently performed using analyte-specific approaches in those areas that practice this, though investigation of genomic approaches is underway [18,19]. Screening of adults for genetic risks, including monogenic or polygenic conditions, is mainly being done currently on a research basis. The specific legal rules regarding genetic and genomic testing vary according to different legal jurisdictions.
Use genomic information to make treatment decisions. • Use and interpretation of pharmacogenetic testing • Targeted pharmacological therapy based on genetic diagnosis • Principles of clinical trials • Application of gene therapy • CRISPR/Cas9 genome editing	Adoption of pharmacogenetic testing has been variable, though the Clinical Pharmacogenetics Implementation Consortium has identified specific drug/gene combinations where testing is recommended [20]. As genes that underlie rare disorders are coming to light, targeted treatments are being developed. Gene replacement or editing remains mostly a research-based effort, though promising results for somatic therapeutic approaches are being reported.
Use genomic information to guide the diagnosis and management of cancer and other disorders involving somatic genetic changes. • Genomic analysis of solid and liquid tumors • Targeted therapeutics and clinical trials • Clonal hematopoiesis • Circulating cell-free DNA analysis	Targeted therapeutics based on sequencing of tumors is rapidly growing in routine clinical practice [17]. Screening for cancer by analysis of cell-free DNA is in investigation as a means of early detection or following response to treatment [21].
Use genomic tests that identify microbial contributors to human health and disease, as well as genomic tests that guide therapeutics in infectious disease. • Genomic tests for infectious disease diagnosis • Role of host factors in disease susceptibility • Role of microbiome in health and disease	Molecular techniques are increasingly used to facilitate the identification of infectious organisms. The role of host factors in disease susceptibility remains an active area of investigation, especially in the context of the COVID-19 pandemic [22]. The importance of the microbiome in health and disease is increasingly recognized [23], though direct clinical application is not yet clear.

- Providing short courses for local practitioners in their native language. The Global Genomic Medicine Collaborative [26] is planning such a course to accompany their meetings in different parts of the world, with the hope that course materials can be repurposed in other educational contexts.
- Development of online course materials in genomic medicine that can be used for asynchronous learning. Systems such as OpenPediatrics [27] could provide a model for this kind of approach.
- Development of a faculty that provides on-site training in different areas of the world using a standardized curriculum.

5 Conclusions

Integration of genomics into medical practice offers new opportunities for prevention, diagnosis, and treatment of disease. The specific needs and opportunities differ in different parts of the world, based on access to technology, priorities in medical care, and ability to provide training. Genomic medicine, however, has moved beyond the stage where it can be limited to the responsibility of a small group of highly but narrowly trained professionals, toward integration across all areas of healthcare. Novel systems of training will be required to customize the ability to prepare the workforce in different parts of the world, taking account of their specific needs.

References

[1] N.J. Timpson, C.M.T. Greenwood, N. Soranzo, D.J. Lawson, J.B. Richards, Genetic architecture: the shape of the genetic contribution to human traits and disease, Nat. Rev. Genet. 19 (2) (2018) 110–124, https://doi.org/10.1038/nrg.2017.101.

[2] P.M. Visscher, N.R. Wray, Q. Zhang, et al., 10 years of GWAS discovery: biology, function, and translation, Am. J. Hum. Genet. 101 (1) (2017) 5–22, https://doi.org/10.1016/j.ajhg.2017.06.005.

[3] A.V. Khera, M. Chaffin, K.G. Aragam, et al., Genome-wide polygenic scores for common diseases identify individuals with risk equivalent to monogenic mutations, Nat. Genet. 50 (9) (2018) 1219–1224, https://doi.org/10.1038/s41588-018-0183-z.

[4] L. Duncan, H. Shen, B. Gelaye, et al., Analysis of polygenic risk score usage and performance in diverse human populations, Nat. Commun. 10 (1) (2019) 3328, https://doi.org/10.1038/s41467-019-11112-0.

[5] B.D. Jenkins, C.G. Fischer, C.A. Polito, et al., The 2019 US medical genetics workforce: a focus on clinical genetics, Genet. Med. 23 (8) (2021) 1458–1464, https://doi.org/10.1038/s41436-021-01162-5.

[6] M. Abacan, L. Alsubaie, K. Barlow-Stewart, et al., The global state of the genetic counseling profession, Eur. J. Hum. Genet. 27 (2) (2019) 183–197, https://doi.org/10.1038/s41431-018-0252-x.

[7] A. Rojas-Martinez, A. Giraldo-Rios, G. Jimenez-Arce, A.F. de Vargas, R. Giugliani, RELAGH—the challenge of having a scientific network in Latin America: an account from the presidents, Genet. Mol. Biol. 37 (1 Suppl) (2014) 305–309, https://doi.org/10.1590/s1415-47572014000200017.

[8] S.A. Vishnopolska, A.G. Turjanski, M. Herrera Pinero, et al., Genetics and genomic medicine in Argentina, Mol. Genet. Genomic Med. 6 (2018) 481–491, https://doi.org/10.1002/mgg3.455.

[9] M.R. Passos-Bueno, D. Bertola, D.D. Horovitz, V.E. de Faria Ferraz, L.A. Brito, Genetics and genomics in Brazil: a promising future, Mol. Genet. Genomic Med. 2 (4) (2014) 280–291, https://doi.org/10.1002/mgg3.95.

[10] Y.M.C. Paz, M.J. Guillen Sacoto, P.E. Leone, Genetics and genomic medicine in Ecuador, Mol. Genet. Genomic Med. 4 (1) (2016) 9–17, https://doi.org/10.1002/mgg3.192.

[11] H. Guio, J.A. Poterico, K.S. Levano, et al., Genetics and genomics in Peru: clinical and research perspective, Mol. Genet. Genomic Med. 6 (6) (2018) 873–886, https://doi.org/10.1002/mgg3.533.

[12] S. Aggarwal, S.R. Phadke, Medical genetics and genomic medicine in India: current status and opportunities ahead, Mol. Genet. Genomic Med. 3 (3) (2015) 160–171, https://doi.org/10.1002/mgg3.150.

[13] B.R. Korf, A.B. Berry, M. Limson, et al., Framework for development of physician competencies in genomic medicine: report of the Competencies Working Group of the Inter-Society Coordinating Committee for Physician Education in Genomics, Genet. Med. 16 (2014) 804–809, https://doi.org/10.1038/gim.2014.35.

[14] S.C. Quinonez, A. Yeshidinber, M.A. Lourie, et al., Introducing medical genetics services in Ethiopia using the MiGene Family History App, Genet. Med. 21 (2) (2019) 451–458, https://doi.org/10.1038/s41436-018-0069-6.

[15] L. Carbone, F. Cariati, L. Sarno, et al., Non-invasive prenatal testing: current perspectives and future challenges, Genes (Basel) 12 (1) (2020), https://doi.org/10.3390/genes12010015.

[16] L.S. Franck, D. Dimmock, C. Hobbs, S.F. Kingsmore, Rapid whole-genome sequencing in critically Ill children: shifting from unease to evidence, education, and equitable implementation, J. Pediatr. 238 (2021) 343, https://doi.org/10.1016/j.jpeds.2021.08.006.

[17] U.N. Lassen, L.E. Makaroff, A. Stenzinger, et al., Precision oncology: a clinical and patient perspective, Future Oncol. 17 (30) (2021) 3995–4009, https://doi.org/10.2217/fon-2021-0688.

[18] S. Pereira, H.S. Smith, L.A. Frankel, et al., Psychosocial effect of newborn genomic sequencing on families in the BabySeq project: a randomized clinical trial, JAMA Pediatr. 175 (2021) 1132–1141, https://doi.org/10.1001/jamapediatrics.2021.2829.

[19] T.S. Roman, S.B. Crowley, M.I. Roche, et al., Genomic sequencing for newborn screening: results of the NC NEXUS project, Am. J. Hum. Genet. 107 (4) (2020) 596–611, https://doi.org/10.1016/j.ajhg.2020.08.001.

[20] M. Relling, T. Klein, R. Gammal, M. Whirl-Carrillo, J. Hoffman, K. Caudle, The clinical pharmacogenetics implementation consortium: 10 years later, Clin. Pharmacol. Ther. 107 (1) (2020) 171–175, https://doi.org/10.1002/cpt.1651.

[21] R.B. Corcoran, B.A. Chabner, Cell-free DNA analysis in cancer, N. Engl. J. Med. 380 (5) (2019) 501–502, https://doi.org/10.1056/NEJMc1816154.

[22] X. Zhang, Y. Tan, Y. Ling, et al., Viral and host factors related to the clinical outcome of COVID-19, Nature 583 (7816) (2020) 437–440, https://doi.org/10.1038/s41586-020-2355-0.

[23] Team NIHHMPA, A review of 10 years of human microbiome research activities at the US National Institutes of Health, Fiscal Years 2007–2016, Microbiome 7 (1) (2019) 31, https://doi.org/10.1186/s40168-019-0620-y.

[24] A. Nisselle, M. Martyn, H. Jordan, et al., Ensuring best practice in genomic education and evaluation: a program logic approach, Front. Genet. 10 (2019) 1057, https://doi.org/10.3389/fgene.2019.01057.

[25] R.L. Wilcox, P.V. Adem, E. Afshinnekoo, et al., The Undergraduate Training in Genomics (UTRIG) initiative: early & active training for physicians in the genomic medicine era, Pers. Med. 15 (3) (2018) 199–208, https://doi.org/10.2217/pme-2017-0077.

[26] G.S. Ginsburg, A global collaborative to advance genomic medicine, Am. J. Hum. Genet. 104 (3) (2019) 407–409, https://doi.org/10.1016/j.ajhg.2019.02.010.

[27] T.A. Wolbrink, N. Kissoon, N. Mirza, J.P. Burns, Building a global, online community of practice: the OPENPediatrics world shared practices video series, Acad. Med. 92 (5) (2017) 676–679, https://doi.org/10.1097/ACM.0000000000001467.

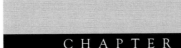

C H A P T E R

9

Digital resources for genomic education and training

Vajira H.W. Dissanayake

Department of Anatomy, Genetics, and Biomedical Informatics, Faculty of Medicine, University of Colombo, Sri Lanka

1 Introduction

The widespread availability of genetic and genomic testing is making it necessary to accelerate capacity development among doctors and other healthcare professionals to provide genetic and genomic services. The situation is further complicated by the rapid discovery of novel genetic variants without the associated clinical data that make reliable clinical interpretation impossible, making it challenging to rapidly translate that knowledge into information useful for clinical decision making at the bed side. There is therefore an urgent need to scale up genetic and genomic education starting at undergraduate education [1]. In almost all countries, genetic and genomic testing as well as education and training for genetics and genomics has developed within the research community and the capacity among clinical staff to deliver education and training in genetics and genomics is limited. There is an urgent need therefore to develop new courses as well as develop innovative ways of delivering them to ensure that the healthcare workforce is educated and trained to deliver genetic and genomic services [2,3]. The COVID-19 pandemic has had two major impacts that those involved in education and training can make use of to further genetic and genomic education and training. They are the renewed interest in basic sciences, especially genetics and genomics, and digital platforms coming to the forefront as the preferred platform to deliver education and training.

2 YouTube as an education platform

Today, YouTube has come to the forefront as an easy-to-access platform for genetic and genomic education and knowledge. Several channels such as Genome TV

Genomic Medicine Skills and Competencies
https://doi.org/10.1016/B978-0-323-98383-9.00010-2

(https://www.youtube.com/user/GenomeTV) and the ACMG Chanel (https://www.youtube.com/user/TheACMGChannel) provide free online access to hundreds of talks by the leading authorities in the field from around the world.

3 Massive open online courses (MOOCs)

MOOC platforms originated as free online course platforms available for anyone to enroll. Today, they have evolved along different business models. However, the courses on the commercial platforms, even the ones from the best universities in the world, remain relatively low cost, making them accessible to a wider group of people from around the world. All MOOC platforms office courses in genetics and genomics. A simple search on MOOC platforms with the keywords "Genetics and Genomics" on November 5, 2021 resulted in 35 courses listed on Coursera (https://www.coursera.org/); 36 courses listed on edX (https://www.edx.org/) and 750 courses listed on Udemy (https://www.udemy.com/).

4 Mobile apps

The global penetration of smart phones has made software application, apps, that can be easily downloaded and installed on mobile devices a preferred mode of delivering not only education but also information/knowledge on demand at the bedside to clinicians and other healthcare workers. A recent systematic review reports the widespread availability of apps [1]. These apps range from those that are made available by professional associations to commercial apps. A comprehensive review of apps is beyond the scope of this chapter. The description below of a selected few highlights their potential not only as educational tools but also as useful tools at the point of care.

4.1 PediaGene app

The American Association of Paediatricians (AAP)'s PediaGene—The AAP Genetics Guide and Screening Toolkit App is a resource containing genetic screening information. The app is based on the AAP Manual on Medical Genetics in Pediatric Practice. Using the app providers can track patients, including images, record family history, see areas of concern, and see screening questions. The app is also a quick reference allowing providers to access and view an image gallery of genetic anomalies, key tables from the manual, and key resources for referral and treatment. Such apps, one would say, are capable of providing continuing medical education on the go, as well as complement patient care at the point of care.

4.2 Face2Gene app

This app is designed to help clinicians diagnose dysmorphic syndromes using facial recognition. It is a useful app at point of care to narrow down and arrive at a syndromic diagnosis.

5 Genomic education and national genomics projects

In the background of the need to increase capacity in genetics and genomics, countries that have been embarking on national genomics projects have made increasing genetics and genomics literacy a key priority of their projects [4]. Among these, the Genomics Education Programme of Health Education England stands out because it is aimed at supporting the mainstreaming of genetics and genomics across the national health service of the entire country (https://www.genomicseducation.hee.nhs.uk/). Online learning is key part of its educational strategy.

6 Challenges and conclusions

The rapid advances in genetics and genomics are creating existing possibilities to transform healthcare. The key challenge, however, is improving genetic and genomic literacy among healthcare workers—many of whom do not have a background in genetics and genomics. The digital tools that we have at our disposal today, which make it possible to access knowledge on demand, and make it available at the point of care, open up new possibilities. The challenge, however, is deploying these tools effectively. These challenges are related to several areas. There is limited knowledge base on the effectiveness of digital education in different settings, utility of acquisition of competencies, and workplace performance of those who were trained on digital platforms compared to those who underwent conventional training. Overall, there is limited knowledge on the impact of digital education, and as such, there is a need for monitoring and evaluation of such programs. The recent development of a minimum set of information to support a consistent reporting of design, development, delivery, and evaluation of genomics education interventions is expected to promote research that would help overcome some of the challenges mentioned above [5]. As digital platforms cross national boundaries, knowledge is on demand anywhere in the world today. Its application in practice, however, may need additional training under competent tutors and educators locally. This also leads to difficulties in regulating professional training and certification of competencies, which in turn would raise questions of professional and legal accountability. These have to be overcome by dialog between regulatory agencies in different countries. There is a dearth of research in this area. We have to acknowledge, however, that digital resources for education and training are here to stay. We have to ensure that these resources are used with the aim of achieving better health outcomes for patients by urgently addressing the issues mentioned above.

References

[1] R.L. Wilcox, P.V. Adem, E. Afshinnekoo, J.B. Atkinson, L.W. Burke, H. Cheung, S. Dasgupta, J. DeLaGarza, L. Joseph, R. LeGallo, M. Lew, C.M. Lockwood, A. Meiss, J. Norman, P. Markwood, H. Rizvi, K.P. Shane-Carson, M.E. Sobel, E. Suarez, L.J. Tafe, J. Wang, R.L. Haspel, The undergraduate training in genomics (UTRIG) initiative: early & active training for physicians in the genomic medicine era, Pers. Med. 15 (3) (2018) 199–208, https://doi.org/10.2217/pme-2017-0077. Epub 2018 May 30. PMID:29843583. PMCID:PMC6008245.

[2] C.K. Rubanovich, C. Cheung, J. Mandel, C.S. Bloss, Physician preparedness for big genomic data: a review of genomic medicine education initiatives in the United States, Hum. Mol. Genet. 27 (R2) (2018) R250–R258, https://doi.org/10.1093/hmg/ddy170. PMID:29750248. PMCID:PMC6061688.

[3] K.A. Calzone, M. Kirk, E. Tonkin, L. Badzek, C. Benjamin, A. Middleton, The global landscape of nursing and genomics, J. Nurs. Scholarsh. 50 (3) (2018) 249–256, https://doi.org/10.1111/jnu.12380. Epub 2018 Apr 2. PMID:29608246. PMCID:PMC5959047.

[4] A.N. Zimani, B. Peterlin, A. Kovanda, Increasing genomic literacy through national genomic projects, Front. Genet. 12 (2021) 693253, https://doi.org/10.3389/fgene.2021.693253. PMID:34456970. PMCID:PMC8387713.

[5] A. Nisselle, M. Janinski, M. Martyn, B. McClaren, N. Kaunein, Reporting Item Standards for Education and its Evaluation in Genomics Expert Group, K. Barlow-Stewart, A. Belcher, J.A. Bernat, S. Best, M. Bishop, J.C. Carroll, M. Cornel, D. VHW, A. Dodds, K. Dunlop, G. Garg, R. Gear, D. Graves, K. Knight, B. Korf, D. Kumar, M. Laurino, A. Ma, J. Maguire, A. Mallett, M. McCarthy, A. McEwen, N. Mulder, C. Patel, C. Quinlan, K. Reed, E.R. Riggs, I. Sinnerbrink, A. Slavotinek, V. Suppiah, B. Terrill, E.S. Tobias, E. Tonkin, S. Trumble, T.M. Wessels, S. Metcalfe, H. Jordan, C. Gaff, Ensuring best practice in genomics education and evaluation: reporting item standards for education and its evaluation in genomics (RISE2 Genomics), Genet. Med. 23 (7) (2021) 1356–1365, https://doi.org/10.1038/s41436-021-01140-x. Epub 2021 Apr 6. PMID:33824503.

Further reading

[6] D. Talwar, Y.L. Yeh, W.J. Chen, et al., Characteristics and quality of genetics and genomics mobile apps: a systematic review, Eur. J. Hum. Genet. 27 (2019) 833–840, https://doi.org/10.1038/s41431-019-0360-2.

Index

Note: Page numbers followed by *f* indicate figures, *t* indicate tables and *b* indicate boxes.